SUSTAINABLE URBANISM

BUILDING CITIES THAT LAST

SUSTAINABLE URBANISM
BUILDING CITIES THAT LAST

Logan Lee
Isabella Gordon

Kruger Brentt
Publishers

2024

Kruger Brentt Publishers UK. LTD.
Company Number 9728962

Regd. Office: 68 St Margarets Road, Edgware, Middlesex HA8 9UU

ISBN: 978-1-78715-118-5

For information on all our publications visit our website at http://krugerbrentt.com/

PREFACE

Urban environmental sustainability encourages revitalisation and transition of urban areas and cities to improve liveability, promote innovation and reduce environmental impacts while maximising economic and social co-benefits. urban development is the way forward for cities to mitigate climate change. Integrated urban places designed to bring people, activities, buildings, and public spaces together, with easy walking and cycling connection between them and near-excellent transit service to the rest of the city. It means inclusive access for all to local and citywide opportunities and resources by the most efficient and healthful combination of mobility modes, at the lowest financial and environmental cost, and with the highest resilience to disruptive events. Inclusive development is an essential foundation for long-term sustainability, equity, shared prosperity, and civil society in cities. Convenience is one of the many reasons people decide to move to a city. People now have simpler access to services like healthcare, education, and others that they might not have had in a rural area. There are many ways to navigate the city rather than staying in the country.

The present book presents the myriad issues of sustainable urbanism in a clear and concise system, and supports holistic thinking about sustainable development in urban environments. It explores four broad measures of urban sustainability that differ radically from other, less long-lived patterns: these are Lifecycle, Aesthetics, Scale, and Technology (LAST). This book is a unique perspective on architecture, urban planning, environmental and urban design, exploring ways for raising quality of life and the standard of living in a new modern era by creating better and more viable places to live. This book will find wide readership particularly among students and young practitioners in architecture and planners.

We are grateful to all those persons as well as various books, manuals, periodicals, magazines, journals etc. that helped in the preparation of this book. In spite of the best efforts, it is possible that some errors may have occurred into the compilation and editing of the book. Further queries, constructive suggestions and criticisms for the improvement of the book are always welcomed and shall be thankfully acknowledged.

Logan Lee

Isabella Gordon

CONTENTS

CHAPTER

INTRODUCTION 1

Cities are built environments and as such they have always been the sites of imaginative and purposeful architectural structures and spaces. Usually they have been constructed to privilege some specific political, ceremonial or religious movement, or to serve some public function, such as a market, or theater, or administrative complex. The great urban centers of Asia, the Middle East and the Mediterranean have long been cherished for the beauty and the architectural imagination and innovation they have exhibited. They are properly seen as evidence of the imagination and intelligence of their designers and builders. Angkor Wat, the Forum, the Acropolis and Chichen Itza are widely celebrated and are the destination of hundreds of thousands of curious tourists and of archeologists and scholars every year. Built with a religious or political objective in mind these and many other similar sites were tied to a region and sought to serve that region in a specific way. When the center was the center of an empire the center was dominant, until its ultimate collapse or abandonment, and other regional centers of the empire were subordinate.

Societies and their cities have evolved dramatically in recent centuries so that the ceremonial and religious aspects of our built environments, while still present, have been eclipsed by other more pedestrian aspects. Perhaps this is because what we value most highly, if not worship, is now more mundane and "practical." Hence, while we still build large religious buildings, rarely are they great or awe-inspiring. Our most distinguished architects now devote their skills to towers of commerce, to indicators of city aspiration, to institutions of cultural display and performance, and to status enhancement, among others. The world economy and society are too extensive and variegated for there to be one imperial center; in its stead we have scores of politically delimited national spaces, each of which has its own values and objectives. This has given us a world not of a dominant imperial center but of many actors each of which is seeking to ensure its own economic survival and competitiveness. To a significant degree architecture has become a tool in this struggle among national economic entities.

Fig. 1. Le Corbusier Radiant City plan

Fig. 2. Central Park in Manhatta

Conversely, public spaces are the absence of buildings other than statues, boathouses and so forth. Often the spaces are created around buildings, as in the Le Corbusier Radiant City plan, while at other times buildings are placed around the space, as in Central Park in Manhattan. While some parks, such as Hyde Park in London, are surrounded by low level buildings, and are designed to bring something of the countryside and nature into the city for the pleasure of residents, Central Park reminds one of a space that was created by stacking living units on top of one another in apartment towers.

In both instances the park is a public space that is designed for repose, recreation, strolling, and admiration of natural beauty. Even small towns have parks, from the New England town commons to waterfront beautification schemes and playing fields of cities and towns of all sizes and importance.

Not all significant architectural works are situated in cities, large and small; many works such as walls, dams and temples are located in rural areas. But in this book we focus specifically on the relationship between the architectural feature and the city in which it is situated. These featured works have a role to play in the life of the city and a function in the city's economic attractiveness. This interplay between city and architecture will be the subject of this book: indeed the use that city leaders make and have made of architecture.

We will find that significant architectural features are a *sine qua non* of the successful city, not just their presence but specifically what they can do to enhance the city and its competitiveness. Again, in the words of Le Corbusier, "Civic pride becomes incarnate in the material achievements of architecture".

THE MODERN CITY

Since the mid-19th century, cities have developed into the principal agents in the global economy. National and sub-national governments are invariably torn between the competing interests and demands of city and rural interests. Cities are also relatively liberal or left-wing in their politics, and this is often in conflict with conservative administrations at the higher levels of government, with the latter being reluctant to transfer funds to the former. The enhanced role of cities is in part the consequence of changes in technology that dramatically altered both the environment in which they existed and their capacity to act in new and more global activities.

City leaders have consciously and purposefully undertaken initiatives to advance the competitiveness of their city in relation to others with which they are in competition for plant locations, jobs, production, exports from the region, and recognition on the part of individuals and firms in other cities. City leaders have utilized a wide variety of measures to achieve these objectives, from public relations efforts to investment in infrastructure to currying favor with higher levels of government for advantages in taxation and regulation.

They have designed metro-wide strategic-economic planning initiatives that have aimed at enhancing the city or in redefining its place in the global economy. In the words of Charles Landry, "For those not in the world league or key secondary cities like Singapore or Sydney, there is intense competition to show distinctiveness. Beyond the cultural and business facilities expected of every large city, the task is to show difference".

It was always thought that the economic game was played by larger cities, and the larger the better. Saskia Sassen wrote of New York, Tokyo and London being global cities, cities that were at the pinnacle of decision-making and command in the global economy. Others wrote of world cities and of global urban regions. Size brought economies of agglomeration, including global hub airports, the full array of financial and professional services, major research universities and an ambience that was attractive to a younger, high technology labor force. All other cities accepted more humble roles, as was appropriate for their lower position in the global urban hierarchy.

Large cities seemed to have it all – dynamic decision-making and economic power in the city center, leafy suburbs for upper income executives with families, more modest housing in the city itself for lower income workers, and ample space on the edge of the city for corporate campuses when the city center was no longer the ideal company site. In recent years, however, the diseconomies of size have begun to emerge – congestion, pollution, lack of public security, impersonality, higher taxes and so forth. Many of the most recent mega-cities, such as the many that have emerged in China, and cities such as Mumbai, Cairo, Jakarta, Delhi and Lagos, are fraught with these problems and with others such as slums, and large populations in extreme poverty.

This brought forth new life for smaller cities. Allen Scott observed that there are numerous small cities in the United States where abundant stocks of cognitive-cultural human capital can be found. These are represented by college towns like Gainesville, Florida; artists' colonies like Taos, New Mexico; tourist centers like Savannah, Georgia; small but dynamic foci of high technology production like Austin, Texas; and so on.

From the first years of the 20th century, authors have analyzed and commented on the essential attributes of urban society. This was at the height of industrialism, with its demands for concentration of economic activity, for suitable labor, for private investment in plant and equipment and public investment in transportation and other infrastructure, and for favorable government regulation and taxation. As we examine the individual aspects in the chapters that follow we will have recourse to the writings of many seminal writers, including Camillo Sitte, Jane Jacobs, Richard Florida, Louis Wirth, Charles Landry, Allen Scott and Ed Glaeser among others, as well as major architects and planners such as Frank Lloyd Wright, Le Corbusier, Mies van der Rohe, Eliel Saarinen and Witold Rybczynski.

Historically, cities had been situated with regard to modes of transportation. At first this meant harbors, large rivers – hopefully with access to the sea – or large inland lakes. One thinks naturally of Athens, Rome, London, Barcelona, Genoa, Geneva, Paris, Stockholm and Copenhagen in Europe, and the large cities of the Atlantic, Pacific and Gulf shores, the cities of the Great Lakes and St. Louis in the US, and Montreal, Toronto, Halifax and Vancouver in Canada. But beginning with the 19th century other locational factors gave rise to cities such as Denver, Dallas, Atlanta, Milan, Berlin and Calgary.

For many it was a new mode of transportation, the railway, trucking and finally air that were important locational factors. All of these cities, and many others, sought to attract production facilities, labor, residents, tourists, financial institutions and other factors that would enhance their economic vitality and competitiveness. One of the most potent tools these cities had at hand was the construction of architectural works that would facilitate their acceptance by other cities as places of significance, as places that must be respected and taken seriously. Unlike Charlie of StarKist tuna fame, there was no confusion between "tasting good" and "good taste" – these cities wanted structures of good taste that would be highly regarded throughout the world.

Industrialization brought with it the great technological advances in materials and structure that have made it possible to realize the architectural works individuals have imagined. It also brought with it increased population density and the need for specialized structures such as housing for workers, office towers, integrated factories, highways, bridges and airports, as well as suburbs, shopping centers, parking facilities and large university campuses. At first it became possible for production units to be concentrated into large integrated facilities such as Ford's River Rouge plant and huge aircraft, steel and chemical factories and districts of the same industry. After World War II, technology advanced in ways that made it possible to break up the large production units and to situate each individual activity in a location that was optimal for it, albeit not for the entire operation.

Geddes referred to this, in 1915, as the shift from "paleotechnic" (18th- and 19th-century industrialization) to "neotechnic" (or modern) industry, although he romanticized it a bit by suggesting this was an evolution from "individual money gains, to the maintenance and evolution of life". Each of these two conceptualizations of industrial activity called forth distinctive architectural projects and also called into question what had suited the previous system: hence the rise and fall of industrial districts and even cities, so dramatically evidenced by the US Industrial Heartland becoming the Rust Belt, and similar experiences in industrial locations throughout Europe and other industrialized countries.

Accompanying this change in economic fundamentals was a change in what we would consider to be the fundamental determinants of urban competitiveness. During the industrialization period cities were more or less competitive in accordance with their endowment of "hard determinants," that is industrial plant and equipment, access to transportation, port and rail facilities, labor with the needed skills, sufficient adequate housing, and access to raw materials. The reader can consider large industrial cities of the 19th century for confirmation of this.

Great industrial centers were created around steel, aluminum, transportation equipment (including aircraft in the 20th century), textile and clothing, and processing of timber, grains and other materials. The endowment of hard assets would correlate closely with the ranking of cities according to competitiveness, although these studies were not conducted until fairly recently.

In the more recent period, heavy industry has given way to financial and professional services, technology-based manufacturing, aerospace, biopharmaceutical and medical technology, and research-based activities. The "hard" determinants were no longer of a dominant importance for a city's competitiveness; they were replaced by "soft determinants." These determinants include aspects that are of importance for the highly demanded and mobile workers in higher technology knowledge-based employment, and include public security, health care, accessible recreation, cultural facilities, high quality education from pre-K through to graduate programs and research universities, and good transportation.

Hence, when the city of today seeks to enhance its economic vitality and its competitiveness, it is these latter assets it wants to enhance, and this has powerful impacts on the sorts of architectural projects it seeks to realize. It must be noted that some of the most competitive cities are small in population but endowed with one or more universities or colleges and a medical complex. We will explore the sorts of projects cities have sought during the 20th and 21st centuries as we go through the chapters of this book.

How does architecture relate to this?

In *Frank Lloyd Wright: In the Realm of Ideas*, Frank Lloyd Wright is quoted as having written: "Architecture is that great living creative spirit which from generation to generation, from age to age, proceeds, persists, creates, according to the nature of man, and his circumstances as they change. That is really architecture". Karsten Harries continued: "'Architecture' names, first of all, the art of building and, second, any structure raised in accordance with the rules of that art".

To put it more mundanely, the substance of architecture can be thought of as: 1) any human-made building or structure, 2) a design for a city, town, park or landscape, 3) the art or science of designing and building buildings, structures, objects and outdoor spaces, and 4) a plan for organizing space.

From "age to age" and from place to place different things are demanded, as will be shown in the description of book chapters that follows this section. Wright's "living creative spirit" is the guiding force behind all of these aspects of architecture. Charles Landry linked this to the project of this book when he wrote: "Creativity has risen because people have realized that the sources of competitiveness now happen on a different plane and they need to learn afresh how to compete beyond merely low cost and high productivity". Patrick Geddes put it this way: "we must ... seek to enter into the spirit of our city, its historic essence and continuous life". Each time and place imposes its own individual demands.

We will not examine the theory or history of architecture, in this book, nor will we focus on techniques of building and construction. We will not be concerned specifically with "great architecture" or with prize-winning buildings; rather our focus will be on the ways in which, over time, architecture has been used to define and redefine a city, both visually and functionally, and to enhance its competitiveness.

Great buildings and other architectural features symbolize the essence of a city, hopefully more truthfully than just the essence that city leaders want to project for it. These architectural projects have impacts, both positive and negative, on the individuals who work and live in them. A city's architecture can attract firms, skilled workers, tourists and entrepreneurs – or it can repel them. This results in architecture being an element in the economic-strategic plan of a city administration. It has been said, perhaps apocryphally, that, in Richard M. Daley's Chicago, the plan for a building would have to be reviewed by the zoning board, a building commissioner, the planning agency and so forth. After all of this, the blueprint would be put on the desk of the mayor and after a while he would declare that "this is a Chicago building" or it "is not." The mayor's overall conceptualization of the city, the aspirations of its residents and how the proposed building contributed to them was all that mattered, for better or worse.

There are many uses we may make of architecture, some of them inspirational, others merely functional, and some responding to rather base motives. The question we must raise here is: What have architects and writers about urban space written that is of benefit to city residents as they try to realize the aspirations they have for their city and for their lives? How have they facilitated achievement of the objectives of city leaders and planners as articulated in the previous section?

We will focus on four ways in which architects have spoken to the issues that confront city economies in the contemporary world. First, the factors that make a city attractive in the modern economy of today differ greatly from those that were important a century ago.

Second, there has developed a powerful need for a city to distinguish or differentiate itself among a world of competitors for plant locations, factors of production and so forth. Third, the importance of city size has been significantly altered in recent decades, with smaller cities gaining in viability. Fourth, technological change has continued unabated and continues to impose its imperatives and possibilities on all aspects of city economic life.

The modern economy has evolved from manufacturing activity done in fixed locations that are endowed with certain hard assets to one of relatively mobile service and technology-dominated activities that give priority to soft determinants. To attract the skilled and mobile workforce, cities need to provide congenial places in which these workers can lead the lives they seek, for themselves and for their families. Both Le Corbusier and Frank Lloyd Wright offered visions of urban space that were largely rejected by the individuals who had to live and work in these spaces.

Le Corbusier sought to create higher density cities of tall buildings separated by green areas, like the apartment areas one sees in Moscow. Creative use of the automobile would make for city centers that were less congested than were the cities that the 20th century inherited. When these principles were applied to Rotterdam they were, rejected by the city residents; when they were imposed on residents who had less control over their lives, principally in the low income housing blocks of major US cities, they were soon dominated by a variety of social pathologies and, as with Pruitt–Igo in St. Louis and the Robert Taylor Homes in Chicago, were demolished.

Wright, too, was taken with the automobile and modern telecommunication and argued that the traditional city was soon to die. His Broadacre City model was composed of one-acre plots of land with room for a house and a garden, connected to the rest of the world by an extensive network of roads, and a space of habitation that could go on endlessly. There was no real city center, but shopping centers were linked by roads and by what decades later became the interstate highway system. Rybczynski sees this as Los Angeles, where Wright lived for a while, writ large. Of course, cities did anything but disappear, and Wright ultimately designed his famous mile high skyscraper for government services in Chicago. Of course, Joel Garreau's Edge City with its corporate campuses and large-lot residential areas is, to a certain degree, a representation of what Wright foresaw.

Quite the opposite of these two architects was the most famous critic of architects and their plans – Jane Jacobs. Jacobs, who defeated Robert Moses in his effort to destroy much of Lower Manhattan with a multi-lane east–west expressway, was famously tied to her Hudson Street neighborhood in Greenwich Village in Manhattan. She was dismissive of virtually everything from urban planning professionals, stating that: "From beginning to end, from Howard (and the Garden City movement) and Burnham (principally his City Beautiful movement) to the latest amendment on urban-renewal law, the entire concoction is irrelevant to the workings of cities. Unstudied, unrespected, cities have served as sacrificial victims".

By ignoring how people actually live and interact, planners and architects such as Le Corbusier, in her opinion, "assumed the statistical reordering of a system of disorganized complexity, solvable mathematically; his towers in the parks were a celebration, in art, of the potency of statistics and the triumph of the mathematical average". She, on the other hand, celebrated the lack of an imposed order and the joy and spontaneity of the human beings who actually inhabited the city, and functioned in the often spontaneous modern economy.

The evolution of cities in the global structure has made it necessary for each city to *distinguish or differentiate* itself in the global context, if only to gain some competitive advantage or stature. A century ago, Geddes wrote of the need to realize the "character and spirit" of an individual city in the emerging "neotechnic" economy. Today it is less the nature of the city that planners want to celebrate, but rather some distinctive feature that will contribute to the attractiveness of the city to economic actors, whether firms or workers, all of which are highly mobile.

One of the most obvious initiatives has been the desire to construct what must at the time be "the world's tallest building." In some places the tallest building has been warranted by economic factors such as the price of land, as was the case with 19th-century Chicago, with its city center confined by Lake Michigan to the east and the Chicago River to the west. "Building up" was a rational solution to a real problem.

Given its geology and the fact that it is an island that is quite limited in its extent, New York was characterized by a series of ever taller buildings that were identified with major US companies, including the New York World Building, the Manhattan Life Insurance Building, the Singer Building, the Metropolitan Life Insurance Company Tower, the Woolworth Building, the Bank of Manhattan Trust Building and the Chrysler Building, each of which claimed title to being the tallest building in the world at the time. Together, over four decades, they helped to establish New York as one of the principal, if not *the* principal, cities of business in the world.

The combining of corporation headquarters and major financial institutions has been a formula for all cities seeking to distinguish themselves through a strategy of seeking to assert their status as centers of economic clout.

A city that is based on high level services rather than manufacturing seems to be a natural place for architects to be hired to design buildings that will give substance to the city's claim for elevated status or differentiation from other cities that are otherwise often very similar. As we will show later, the Petronas Towers did this for Kuala Lumpur, as did the Burj Khalifa for Dubai. Neither of these cities was seen as a principal economic center until these structures brought them to the world's attention.

In the previous section we quoted Scott on the importance of city size and on the attractiveness of smaller cities, cities other than the global and world and mega-cities that have captured so much attention. Here we will discuss the relation between the spatial or physical size of a city and the work of architects and the practice of architecture. Some have argued that cities should be very low density places lacking a strong center and with residences and places of work and production spread out over a great space. This was the approach of Frank Lloyd Wright in his Broadacre City with its Usonian home and its Architecture of Democracy.

The use of the automobile and of telecommunications made this conception of a city conceivable, and was, for Wright, a way of maintaining democracy, a relatively egalitarian distribution of income, in the political structure of social democracy.

Corbusier used the automobile and telecommunications to realize an alternative end, that of a large number of very tall buildings separated spatially by extensive green areas – in his plan the buildings would be placed on 15 percent of the land and the rest would be planted. Beyond them lower structures would extend out to the fields where food was grown. Both of these schemes would generate what is now derogatorily referred to as "sprawl."

Sprawl is alternatively seen as a situation in which individuals can all realize a Wrightian home and garden on one acre of land, among other things, or conversely as the leapfrogging of empty spaces to build on cheaper land farther from the city center, which generates enormous diseconomies and waste in terms of time to do anything other than swim in one's own pool and of energy used in transportation and in inefficient single dwellings.

At the other end we have the extensive groupings of high rise apartment and office buildings that dominate places such as Manhattan where land is physically limited and there is no alternative to building up, or the many places in China where clusters of tall residential buildings are situated seemingly with more regard for the availability of zoning permits and financing than for the availability of potential residents. Many of these tall structures are quite magnificent and contribute much to avoiding the creation of sprawl. It is in these denser aggregations that one finds the great cultural institutions, such as museums, concert halls, theaters, and the creativity that occurs in environments that offer opportunities for the much vaunted face-to-face contact and transfer of tacit knowledge. Architects have responded marvelously to the need for buildings to house these cultural activities in virtually every city of consequence.

Technological change has enabled great architects to imagine and then to construct the buildings that have contributed so much to the attractiveness and vitality of our cities. Buildings went from Burnham & Root's Monadnock in Chicago, in 1891, with its 12-foot thick at the base walls (needed to support the weight of the ten stories of masonry walls), to their steel skeleton frame glass Reliance Building and Jenney's Home Insurance Building (the first skyscraper) four years later, to the Burj Khalifa, in Dubai, with 183 floors and a height of 2722 feet. Projected for completion in 2020 is the Kingdom Tower in Jeddah with a height of over 3300 feet and 252 usable floors. All this happened in just 130 years, thanks to the ingenuity of the variety of professionals involved in this development. Steel framing was the first major innovation, as it allowed for the extreme height of buildings that was not possible with stone, brick or iron.

Fundamental to all of this were the invention and improvement of the elevator, electricity for implements and for lighting, thermo-glass panes for windows of all sizes and shapes, advances in the uses of computer technology to solve problems from calculating the forces of wind vortices on tall buildings to designing the undulating sheets of a Frank Gehry concert hall, and advanced telecommunications to allow the spatial separation of workers on the same project, whether vertically on different floors in the same building or horizontally across town or in between towns. Finally, subways and other mass transit had to be developed to get people to and from the building in an efficient manner.

Technological advance has enabled architects to realize the buildings they have imagined. Frank Lloyd Wright's Mile High Illinois Building would have had 528 floors and lower structures hundreds of feet below the surface.

It was immediately thought to be a fantasy building with no chance of being constructed. But the Kingdom Tower will be two-thirds that height, and this would have been thought impossible a few years ago.

Changes in the design of tall structures have enabled planners to conceive of vertical cities in which certain efficiencies and economies can be achieved by an extreme exaggeration of the Corbusian notion of tall buildings nested in green spaces. It has become legitimate to ask: What are the technological limitations to the height of buildings? The development of carbon-fiber rope has meant that elevators in tall buildings can travel to greater height, obviating the need to change from one elevator to another after a few dozen floors. The original motive for tall buildings, as in 19th-century Chicago, was the cost of land and the availability of prime land. Vertical cities with clusters of 200-story buildings linked by skyways seek to capture a different set of efficiencies and economies. And many of these tall structures are being built in cities in which the climate does not invite a leisurely stroll through the street or a park. Trefil imagines one moving into the Hancock Tower in Chicago upon graduation from university and not having to leave the building, with its shops, gyms, theatres and so on, until he or she retires. Not that this would be an inviting prospect for all of us!

CONSTRUCTING THE CITY OF THE FUTURE

It's hard to imagine cities looking much different than they do today. Towering skyscrapers, bustling sidewalks and busy traffic have been the norm for decades, and tightly-packed buildings leave little room for new construction. However, as the global population grows and more people flock to urban areas, new methods to build efficient, sustainable cities could drastically change the urban landscape in just a few decades.

Cities all over the world are searching for new ways to expand urban landscapes, reduce pollution, and create safer, healthier spaces for their residents. The world's future cities will rely on advanced construction technology and innovative urban planning initiatives to achieve this, some of which are already in development today.

Read on to learn how construction technology and urban planning trends are paving the way for future cities or skip to the animated infographic below to see what cities could look like in 50 years.

CONSTRUCTION TECHNOLOGY PAVING THE WAY

As plans for the city of the future begin to take shape, so do the construction techniques required to build it. The future of urban construction will rely on cutting-edge technologies that redefine how buildings are built and add a level of efficiency never seen before.

3D MODELING

Construction companies are already beginning to take advantage of these technologies to create safer, more efficient structures. Architects and engineers can use building information modeling (BIM), an intelligent software modeling process, and artificial intelligence (AI) to view 3D building designs and make adjustments before construction even starts. These technologies can also be used to monitor progress throughout the construction process and catch mistakes before they happen.

BUILDING AUTOMATION TECHNOLOGY

The green building market is expected to reach $99.8 billion by 2023, and it will only continue to grow as governments crack down on CO2 emissions. Many future cities will rely on eco-friendly construction to meet their sustainability goals, and building automation can help achieve that.

Building automation systems are relatively common in the construction industry, but companies are just beginning to use them to regulate energy usage and reduce waste. Automated HVAC, electrical and mechanical systems can regulate energy consumption and storage and eliminate human error. Some buildings are already saving up to 30% on energy costs by utilizing automation, and it can help future buildings become green-certified.

PREFABRICATED BUILDING MATERIALS

Modular construction and 3D-printed building materials are also great candidates for reducing the environmental impact of future construction projects. These technologies allow builders to create sustainable materials offsite and help reduce construction waste. 3D-printed building elements can be constructed using durable materials like sand, metal, and concrete, and they require fewer workers to construct.

WHAT WILL FUTURE CITIES LOOK LIKE?

Over the next few decades, cities could look very different than they do today. Many future advancements are already in the planning and development stages, and they could become the norm in as little as 20 to 50 years. The following developments give a taste of what future cities could look like.

Modular Skyscrapers

Modular construction is a small but growing subset of the construction industry that aims to streamline the building process and promote sustainability. Modular technology is often used for homes and smaller buildings, but modular skyscrapers are becoming popular in urban areas that have rapidly growing populations.

Modular skyscrapers are built with prefabricated units, or modules, made from sustainable building materials. These modules are constructed off-site, then transported to the construction site where they are quickly assembled. This process condenses the project timeline and reduces construction waste, making it much more efficient than traditional building methods.

The world's tallest modular skyscrapers will soon open their doors in Singapore, and more are soon to follow. The Avenue South Residences are two 56-story residential towers made from 2,984 prefabricated modules.

After construction began in 2019, architects noticed that the project required fewer on-site workers and allowed factory employees to socially distance more than they would on a busy construction site. New social distancing standards and altered workplace models put in place during the COVID-19 pandemic could accelerate the push for modular buildings and create new demand for socially distanced construction methods.

Net-Zero Energy Buildings

Experts predict that 70% of the world's population will live in urban centers by 2030. Many cities are already struggling with dangerous levels of pollution and carbon emissions, and growing populations will only add to the problem unless urban planners find more sustainable solutions. Fortunately, net-zero energy buildings offer a green alternative to traditional building models.

To keep global warming below two degrees Celsius, countries that signed the Paris Agreement will need to create a net-zero building sector by 2050. Cities can achieve this by building net-zero energy commercial and residential buildings that use renewable energy sources and only use as much energy as they can create on-site. Net-zero energy buildings have yet to be adopted on a global scale, but national and state-level emissions mandates have urged governments and private organizations to invest in the technology. For example, the National Renewable Energy Laboratory is currently researching new renewable energy technologies, and California's recent climate change legislation calls for all new construction projects to achieve net-zero emissions.

Smart Highways

Highway systems have remained relatively unchanged over the past few decades. However, with electric vehicles entering the market and self-driving cars on the horizon, our roads and highways are due for an upgrade.

Companies all over the world are starting to invest in technologies to make highways more efficient, safe and sustainable. Multiple technologies can be incorporated into a roadway to make it "smart." Some of the technologies currently in development and planning stages include:

- **Temperature Sensitive Road Paint:** Adding photoluminescent powder to road paint causes road lines, barricades and other markings to light up at night, making roads safer to navigate.
- **Weather, Traffic and Maintenance Sensors:** Dynamic sensors in highway systems can detect and transmit data on weather conditions, repairs and traffic delays.
- **EV-Charging Roads:** Roads embedded with electric rails and other hardware could charge electric vehicles as they drive.
- **Solar Roads:** Solar panels placed along roadways can power illuminated street signs and heat strips to help melt snow and ice.

Some examples of smart highway technology have already hit the road, including automatic electric tolls, motion sensors on traffic lights and GPS technology that tracks your location as you move. More advanced highway technology might feel far off, but countries like Sweden and the Netherlands are already testing out EV charging roads and glow-in-the-dark road paint to streamline their highway systems.

ELECTRIC PUBLIC TRANSIT

- Public transportation has always been a greener alternative to cars, cities can take it a step further by implementing fully electric public transit systems that reduce CO_2 emissions and eliminate the need for fossil fuels.
- The U.S. introduced the BUILD GREEN Infrastructure and Jobs Act in March 2021 as part of a plan to electrify the country's public transportation systems. This bill would put $500 billion towards electrifying the nation's trains and buses and support the current administrations' plan to transition to a renewable energy future. The US Department of Transportation also recently began awarding a total of $182 million in grants to transit agencies wishing to add electric buses to their fleets.
- This push for electric public transportation is happening in other countries as well. The European electric bus market rose 48% in 2018, and China currently has the largest fleet of electric busses in the world. A Bloomberg

report predicted that half of the world's buses could be fully electric by 2025, making electric transportation a dominant force in cities of the future.

FUTURE URBAN PLANNING TRENDS

Urban planning trends change over time as cities grow and evolve. What works right now might not work 20 years down the road, and urban planners are constantly searching for new ways to update outdated systems. The following urban planning trends are still in their infancy today but could become the norm before we know it.

15-Minute Cities

Fueled by climate change and the COVID-19 pandemic, 15-minute city concepts are quickly gaining traction among urban planners across the globe. 15-minute cities are communities in which everything a resident needs—including work, food, healthcare and entertainment—can be reached in 15 minutes or less by foot or bike.

While the idea has been around for decades, the modern 15-minute city concept originated in Paris, France, a city known for its frantic, fast-paced energy. Anne Hidalgo, Paris' mayor, appointed a commissioner for the 15-minute city and has made the concept a top priority for future urban planning projects. Some of her recent projects include creating mini green spaces throughout the city, banning high-polluting vehicles and banning motorized vehicles from the Seine Quays.

15-minute cities are gaining momentum around the world as virtual and hybrid workplaces become more common. As people begin spending more time in their communities, the demand for neighborhoods with a more diverse range of businesses and public resources will continue to rise.

BLUE ZONES

The term "blue zones" was first coined by author Dan Buettner in 2005 after he studied communities around the world where people lived longer than average lives. Blue zones are regions where the world's oldest populations live, including Okinawa, Japan; Sardinia, Italy; Nicoya, Costa Rica; Icaria, Italy; and the Seventh-Day Adventist community in Loma Linda, California.

Blue zone inhabitants experience fewer instances of chronic disease, and they live healthier lives on average than the rest of the world. Buettner attributed this to the following four concepts:

- Living with purpose
- Moving regularly
- Robust social support systems
- Making the "healthy choice the easy choice."

While the concept of blue zones is heavily tied to human behavior, it can be applied to urban planning on a citywide scale. Fort Worth, Texas and North Port, Florida have already moved to adopt blue zone principles by creating walkable environments, community and rooftop gardens, local gathering spots and a mixture of development and housing types.

ECODISTRICTS

Ecodistricts are neighborhoods that limit their environmental impact by prioritizing sustainable development and reducing energy consumption. Similar to blue zones, ecodistricts are also designed to promote diverse social environments to increase residents' quality of life.

They achieve this by utilizing sustainable and recycled building materials, renewable energy sources, eco-friendly forms of transportation and state-of-the-art waste management systems.

While ecodistricts are popping up in cities all over the world, countries in Northern Europe—like Sweden, Germany and France—are leading the effort. The city of Paris alone has at least four ecodistricts, and plans for more are currently in the works.

The Hammarby Sjostad ecodistrict in Sweden equips its buildings with solar panels to provide electricity, and its heating system is fueled by biofuel, combustible waste and water from its water treatment plant.

MODERN CITIES LEADING THE WAY

Some cities are already adopting futuristic urban planning trends to kickstart their transition to a more sustainable, efficient future. The following cities have implemented unique technologies that create a model for what cities of the future could look like.

Los Angeles, California

Los Angeles is at the forefront of sustainable urban innovation in the U.S. Because LA is a coastal city in a state prone to drought, urban planners have kickstarted efforts to make the city more sustainable and drought-resistant in recent years. The city adopted a Green New Deal in 2019 as part of its sustainable city plan. It set the following targets:

- 100% renewable energy by 2045
- Recycle 100% of wastewater by 2035
- Reach 100% net-zero emissions by 2050
- Create 400,000 green jobs by 2050

- Fully electrify transportation by 2030
- Divert 100% of waste from landfills by 2050

Fig. 3. Los Angeles, California

To help achieve these goals, the city has already incorporated sustainable architecture into its urban planning initiatives. Los Angeles became the first city to control its street lighting through mobile and cloud-based technology, ensuring that lights aren't left on longer than they need to be. Landscape architects have also begun replacing water-inefficient plants with native, drought-resistant flora in urban areas to help conserve water and diversify the landscape.

La Paz, Bolivia

At up to 3,600 feet above sea level, La Paz is one of the highest cities in the world. The city is known for its winding streets and steep mountain slopes, making it simultaneously beautiful and difficult to traverse. For those who don't have access to a car, getting around used to require setting out on foot or relying on the city's minubusses, a fleet of privately-owned vans with unpredictable routes.

To revolutionize its transportation system and connect previously inaccessible parts of the city, La Paz began building an advanced cable car system called "Mi Teleférico" in 2014. The system consists of constantly moving cable cars that can fit up to 10 people and have almost no wait time. Since its inception, Mi Teleférico has expanded to include seven operational lines stretching from La Paz to the city of El Alto, a growing metropolitan area bordering the capital city. Mi Teleférico is the first public transportation system in the city that supports the needs of people with disabilities and impaired mobility. Its stations span across multiple socioeconomic urban areas, and it significantly reduces commute times for workers traveling from El Alto to La Paz. It also only costs three bolivianos ($0.45) to ride, making it accessible to low-income commuters.

Mecca, Saudi Arabia

Mecca, Saudi Arabia is home to the Kaaba, one of the most sacred sites of the Islamic faith. Around three million Muslims travel to Mecca every year over five days to visit the Kaaba, and the city has had to adapt to the massive influx of people and overwhelming crowds.

The Saudi government invested over $20 billion to upgrade the Masjid al-Haram, the Great Mosque, to accommodate the throngs of people that travel there every year. This effort involved building eight-sided platforms around the Kaaba, a holy site that all travelers visit on their hajj trip, to help streamline the flow of people.

Engineers also installed a high-tech series of pneumatic tubes at the Great Mosque to accommodate the 600 tons of trash created on-site each day. The trash is dropped into openings on-site, then sucked at 40 mph through an underground system to a station over a mile away.

Today's most advanced cities and urban planning trends could lay the groundwork for city planning for years to come. While we can't predict exactly what the city of the future will look like, it will reimagine the way we think about sustainability, transportation and the way buildings are built.

CHAPTER

2 SUSTAINABILITY, RESILIENCE AND PLACE IDENTITY IN FUTURE

While the advent of cars was celebrated for the numerous possibilities it brought in terms of mobility, trade and social class association, it has had numerous documented negative and severe impacts on the socio-economic fabric. In cities, cars changed the dynamics of urban planning, opening doors for linear and perpendicular city grids and the devastating consequences of urban sprawl. The wide-ranging availability of cars flooding streets in cities, as noted by researcher, has led to continuous negative impacts on both biodiversity and quality of life, linked to negative outcomes such as increased traffic congestions, which, besides psychological and social tolls, translate to time and economic loss. Such outcomes are also linked to low air quality as a result of increased emissions and the energy demand (fuel) to keep vehicles on the roads.

Appleyard interestingly notes that the availability of cars has also resulted in increasing financial burdens on different households, resulting from the lost time in traffic, the demand for fuel and the increasing cost of vehicle maintenance. Regarding biodiversity, cars have led to urban sprawl, negatively impacting green reserves and forest and arable lands as more people opt to reside in those areas to escape chaotic urban environments; entering a catch-22 situation where the need for habitable environments need to be balanced with new infrastructures tied to accessibility to basic amenities, which are not always observed in sprawled areas.

This form of urban approach could be traced back to the celebrated modernist architect le Corbusier, who laid the blueprint for urban sprawl in his bid to escape the unsustainable urban environment. He supported that cities were not efficiently planned for them to be considered "sustainable", and the only alternative was to demolish and re-plan and rebuild with sustainable dimensions being infused. However, his immediate proposed solution promoted urban sprawl and has, over time, proven distasteful—and coined by Jane Jacobs as being "the most dramatic idea"—as it encouraged excessive reliance on automobiles. The creation of urban environments accommodating cars has prompted the demand for vehicular-dependent cities, and this has, over time, been satisfied through the mass production of automobiles, which paradoxically aligned with the global agenda of industrialization and oil dependency, which have been later understood as being among the epoch where emissions escalated the impacts of climate change.

Today, our car-dependent urban planning legacy outlines deep-rooted inequalities, especially in the social and economic spheres, and has become the center for unsustainable practices. However, in some major cities, there are some efforts to revamp and encourage public transit mechanisms such as the use of public transport with the objective of reducing private car ownership and emissions.

However, before innovations such as the use of virtual communication options, cities that were partially or totally locked down experienced unprecedented challenges such as the shortage of basic supplies such as food. A sizeable number of urban residents also faced unemployment as companies and institutions downscaled due to the low demand and supply of basic materials and services. The emergence of this pandemic exposed the vulnerability of cities in their current establishment and the need for a radical re-thinking, where innovative measures need to be tailored to ensure that urban residents are able to cope and continue with their basic activities, including cultural ones, to ensure that cities remain both resilient and livable in the short and long terms.

The need for this radical re-think, coupled with pressing socio-economic issues, led cities to adopt urban planning mechanisms to ensure that quality of life is preserved beyond the fractures that the virus will possibly leave as legacy in cities. On this note, policy makers view it as a necessity to provide proximity-based services to inhabitants by reviewing urban policies, particularly transportation, viewed as a connective thread to solving other issues. In view of this background, we propose to study a conceptual approach that has captured the attention of many politicians around the world: the "15-Minute City", as imagined by the first author, Carlos Moreno, the implementation of which was accelerated during the COVID-19 pandemic.

With this backdrop, we map recent urban trends pointing to the "15-Minute City" concept and analyze similar approaches that insist on the concept of "chrono-urbanism" as a way to improve quality of life in cities—the 15-minute walkable neighborhood proposed by Weng et al. and the 20-Minute City model, advanced by Da Silva et al.—as viable models for modern cities to address their underlying urban challenges. Through this approach, we explore the "15-Minute City" concept, introduced by Carlos Moreno, and discuss its merits and areas for future research.

This perspective section introduces the 15-Minute City model as an alternative planning approach in light of the global pandemic. This perspective is compared with other emerging, yet similar, planning models, where the differences are mapped and explained in , while the fifth section comprehensively introduces and explains the proposed concept, including four distinctive dimensions pertinent to the proposed planning model.

A PERSPECTIVE ON THE 15-MINUTE CITY AS AN URBAN PLANNING PANDEMIC RESPONSE

Since it was first reported, the impacts of the COVID-19 pandemic are still unfolding, with numerous countries grappling with a second wave of the virus outbreak, reported as being more lethal than the first one. Due to this, most countries were contemplating the re-introduction of the already eased containment measures to mitigate further spread of the virus.

The re-introduction of stringent health protocols and lockdown measures is expected to further dent the economic and social dimensions of cities, which were trying to rise from the shock of the first brutal wave, especially on urban dwellers. The situation could have been complicated by the modernist urban planning concepts that created fragmented cities and communities, being heavily automobile-dependent. With this legacy, most urban fabrics, though connected, are not easily accessible without a heavy reliance on automobiles.

Such a situation had caused bottlenecks in service delivery and calls for the need for more appropriate urban planning mechanisms as a means of pandemic response in different cities. Going further, Allam and Jones called for the need for post-pandemic urban planning mechanisms to dwell into urban resilience for higher livability dimensions.

One such temporary urban response that has proven efficient in lockdown cities, where vehicular use had drastically declined, was the inclusion of bicycle lanes to counter the impacts arising from vehicular movement restrictions.

De Vos notes that encouraging bicycle use during this pandemic has had numerous benefits including on the health and well-being but also in helping people maintain social contact while observing health protocols and adhering to the long-term goals of sustainability.

The influence of bicycles on promoting sustainable mobility and COVID-19 abatement is evident in cities such as Berlin in Germany, where already, when the pandemic broke, two temporary bike lanes were set up, followed by the pursuance of long-term policies on the subject.

Another example is in New York City, which witnessed an upsurge in bicycle usage by over 67% in March, where its success led to calls for permanent transformation to become a "bicycle city". In Beijing, bicycle usage also increased by over 150% as people tried to avoid public transport and comply with the restrictions on private car usage. In Copenhagen, bicycle shops were considered essential service providers and were never closed, even when other non-essential service providers were forced to. Adding to this, other services linked to public spaces, parks and cultural services and amenities—often overlooked in the past—saw their popularity and demand rising during the pandemic.

The world has been amassed with innovative temporary urban infrastructures, which has helped to partially deal with some challenges presented by the pandemic. Such success highlights that the public wishes that urban managers will concentrate henceforth on investing in proximity-based needs rather than rolling out further investments in large infrastructural projects as a means of economic support through relief packages. For instance, in the Netherlands, micro-markets—created by Timmermans—helped to ease congestion in supermarkets, proving essential in ensuring that locals were able to access basic commodities within their close proximity, hence reducing the risk of infection in so doing. Another temporary infrastructure is a maze-like park, introduced in Austria to help locals experience green spaces as they maintain social distancing measures. The architectural design of the parks made them implementable in any open space, maximizing, in effect, the use of unused public spaces in cities, especially those challenged with land scarcity for establishing public parks.

Such temporary infrastructures are just an example of the many that could be implemented in the pursuit of the "15-Minute City" concept. The success of this concept, as it has been shown in the city of Paris under the leadership of Mayor Anne Hidalgo, has been hailed as a potent urban planning concept that will lead to an economic boost, while bringing about social cohesion and interaction and help create sustainable ecosystems in cities, more so after the experiences of COVID-19 and associated containment measures.

While some of the features of the "15-Minute City" concept had been temporarily adopted in different cities after the impacts of COVID-19, its adoption in long-term planning would result in a higher quality of life as proximity to basic services would help in saving time wasted in traffic, thus promoting sustainable mobility. This will aid in efforts to reduce emissions as envisioned in the Paris agreement and promote higher cultural outputs, amongst others.

For instance, by re-thinking the transportation system to create more biking and walkable streets, the challenges of private car ownership will be somehow addressed as they will be reduced as more people embrace biking culture. In addition, as expressed by Reimer, the adoption of the "15-Minute City" concept will also open gateways for more novel digital innovations such as bike-sharing technologies that would increase the high livability experiences of urban residents. For instance, as is expressed by Gehl, the re-thinking of cities to facilitate walkability and cycling would, in turn, inspire the creation of parks, squares and public places within neighborhoods, and by doing so, it would help to bridge the social inequality in accessing such facilities, which are not always available for everyone in a car-dependent city.

AN EARLY MAPPING OF THE 15-MINUTE CITY CONCEPT AND EMERGING VARIATIONS

Since the concept's proposal in 2016 and its wide coverage in the popular media, followed by its adoption at policy level in Paris, leading to discussions in other global cities, there have been emerging variations of the concept that seem to portray the same principle philosophy of "chrono-urbanism" while supporting the need for proximity-based indicators to better service urban areas. These approaches of chrono-urbanism are in line with studies underlining the importance of urban rhythms in order to understand the quality of life in cities: especially space is relevant only as it is coupled with temporal dimension.

The 15-Minute Walkable Neighbourhoods

Weng, et al. explore the significance of 15-min walkable neigborhoods that cater for the need of all the demographics. In particular, the authors promote this as a way of promoting the health dimensions of the residents especially in voiding non-communicable diseases like obesity. However, the authors were concerned that the 15-minute walkable neigborhoods could present some form of social inequality in relation to attributes such as age and economic status.

That is, they opined that the basic amenities that younger people like children may be interested in (such as sport facilities and playgrounds) may be different from those elderly people may be interested in and noting that such different groups have different walking behaviors and speed, urban planners need to be informed. However, the authors observed that within walkable distance, the social needs of different demographic groups were not sufficiently provided, and at any given time, one or more group was left disadvantaged in provision of basic amenities.

As advanced in the "15-Minute City" concept, the authors are of the opinion that basic urban amenities need to be provided in close proximity to urban centres, and such should not disadvantage people in respect of their socioeconomic status or age. This article is also inspired by the perceived need to maintain close social ties for increased liveability indexes. On this, it is noted that in most cities, including Shanghai which is the case study for the article, basic amenities are located in the central areas while the peripheries where residential areas are located have those amenities sparsely distributed; thus, making walkability almost impossible.

Philosophically, the authors view the planning of Shanghai city with the same lens as Le Corbusier, who preferred a city that seem to be strongly ordered, but which ultimately necessitated the need for automobiles to access all the available amenities.

While this section principally correlates with the concept by Moreno, it differs on how it emphasizes the need for walkable neigborhoods as a way of promoting a healthy lifestyle. Furthermore, the need for a 15-Minute City is equally focused on other dimensions, relating to ecological sustainability, promoting social interactions and citizen's participation and addressing automobile dependence by emphasizing on proximity of all basic services.

The 20-Minute City

Another variation to the "15-minute city" concept is by Capasso Da Silva, et al. who note that it is possible to plan cities that are accessible within a 20-Minute walk, cycling or transit. The authors' conviction is that during the planning stages, concentration should be on accessibility rather than transportation aspects which have inspired most cities to date. In the section, the city of Tempe, Arizona has been forwarded as the case study, and the authors express that though the city is accessible within a 20-minute of cycling or transit, it is experiencing some form of sprawl that would eventually dilute this concept of "20-Minute City". That is, with the current trends of urban sprawl, the concept would only hold if residents depended on cars to access basic essentials.

Though the section advocates for a "20-Minute City", it aligns with Carlos Moreno's concept of "15-Minute City", especially on how it emphasizes on accessibility and the need to reduce mobility. The authors are emphatic on the need for cities to consider their built environment to allow for accessibility in the city. According to them, planning emphasis need to focus on ensuring streets in urban areas are organised in such a way that they promote accessibility to different parts of the city within 20 Minute walking radii, and such planning would sometimes require restricting or reorientation existing infrastructures

Apart from accessibility, this section differs with the concept of Moreno in that it does not emphasize on need for sustainable social interactions and participation of urban residents. There also seems to be a lack of emphasize on the need for proximity of basic amenities and services, and instead, its concentration is on accessibility to opportunities, which in this case are aligned to job opportunities; an interesting dimension that strongly relates to automobile usage.

While the emphasize on access to opportunities is novel, such does not expressly highlight how such accessibility would help in minimising reliance on automobiles by residents in their search for services like parks, playgrounds, schools, shopping centres and others. Another philosophical divergence that these authors have with the "15-Minute City" concept by Moreno is that they seem to advocate for a different distance based indicator. This means that, though the basic amenities would still be available, in the 20-minute city, they would be a bit further placed than in the concept advocated by Moreno. It would also mean that much land and more resources will be required to actualise this planning model. However, beside those divergences, the ultimate objective in the two concepts are similar, with proximity, and reduction of car reliance being some of those objectives.

ON SMART CITIES, COLLECTIVE INTELLIGENCE AND PLACE IDENTITY

The"15-Minute City" concept, further expounded on in below, aligns with the tenet of "chrono-urbanism" and advocates for increased proximity, social interaction captured by the "density" dimension, digitalisation and diversity pillars which in essence would lead to more closely knitted community fabrics. This would by a larger margin be facilitated by the increased technological advancements that have led to the emergence of such novel urban planning models like the Smart City concept, which is based on the success of digital revolution. On this, the underlying principle of Smart City concept is the deployment of diverse technologies to optimize different urban fabrics. Numerous authors champion this concept as a novel model that can assist cities to achieve sustainability and improved liveability status for an increased quality of life for their residents.

The achievement of Smart City concept in the current epoch is tied to availability of a rich interconnected array of Internet of Things (IoT) devices that have the capacity to collect, and send data to a centralized network, where this data is then computed, analyzed and distributed, in real-time to respective urban department for actions. The availability of IoT devices, which Alam note will be in excess of over 75 billion devices by 2025, coupled with technologies such as Artificial intelligence (AI), Big Data, Machine Learning and Crowd Computing and others is expected to actualize the proposed 15-Minute City concept. At the core of the concept is the availability of different amenities and urban services; which with technology, would render it easier to use available data on different aspect to ensure novel urban planning to ensure optimal service delivery within intended areas of proximity.

Aligning with calls of technological integration in cities to achieve the dimensions outlined in the Sustainable Development Goal (SDG) 11, pertaining to safety, sustainability, resilience and inclusivity, the use of diverse technologies, smart devices and sensors can lead to this actualisation and in effect render a higher quality of life.

It is further observed that improving the quality of life of urban residents leads to a focus on four major converging aspects: (1) promoting social inclusion for better social cohesion and combating exclusion, (2) reinventing urban infrastructure to adapt to 21st century lifestyle changes, (3) building on the digital revolution, (4) taking into account major urban environmental issues, including the challenges of climate change.

In this regard, different aspects of the urban fabric are interlinked using an array of IoT devices, generating a rich database, critical in providing invaluable insights shaping policies and projects. As noted by Ferrero Bermejo, et al., with the data, cities pursuing "smartness" are able to address challenges like increasing energy demand, traffic issues, social inequality in housing and in provision of services like health, improve economic status of residents and confront the challenge of sustainability heads-on. In reference to the 15-Minute City concept, technology use is thus novel, especially in ensuring that alternative solutions -aligning more with the tenets of the SDG 11, are available in diverse areas, including that of the transport sector in order to repair cities from the decades of fracture that car-driven policies left as legacy. For instance, with technology such as AI, IoT and Big Data, technologies such as bike sharing, online shopping, car sharing, and drone delivery services (currently being tested by Amazon), amongst others, are on the rise in different cities. Such services are geared to ensure that people are safe and on time by eliminating the need to travel to purchase different items, and where they have to travel, the use of bicycles, or walking ensures they experience no traffic and are able to safe on costs that would have gone to autos. However, technology here, should not only be about removing the need to access and engage in the primary activities necessary to sustain an urban life, but also aid -through the use of policies, to ensure that a high quality of life is achieved while doing so.

In respect to the ecological aspects promoted in the "15-Minute City" concept, modern technologies have been identified as being a proponent in the prospects of achieving sustainability and resilience in cities. For instance, one area that Smart City researchers advocate for is the adoption of mixed-use housing models that promote compactness, efficient use of resources like energy and water, increase in proximity and in the promotion of adequate land use. When compounded, all these factors have a positive bearing in promoting ecological sustainability, while at the same time enhancing the quality of life of the citizens. Further, by adopting modern technology, especially in the building, planning and construction sectors, modern trends being pursued in the concept encourage the incorporating of green and civic aspects in building and planning. That is, encouraging the use of green building and planning systems and technologies, including green spaces in the built environment, embracing use of green roofing and walls among others -through the careful planning, optimisation and retrofitting of spaces and built structures.

Adding to this philosophy, we advocate for the need for quality human interaction and that participatory processes are encouraged at varying scales of planning processes, through conception to after construction, through feedback loops. With modern smart technologies that cities are pursuing, this aspect of interaction is becoming possible, and should become a norm as is widely expressed by Gehl. For instance, in respect to participation in a diverse society, it has been observed that urban managers and planners are now able to interact with residents using different communication platforms through both closed (through intranet systems) and open loop system, like social media. The latter have gained popularity during the last decades since the advent of social media platforms, coupled with the decreasing price of internet connectivity and increasing bandwidth rate (from 2 Mbps in 2000 to over 1 Gbps in 2015). For the internet, the NCTA notes that the speed increased by over 10,000% fold in the last 15 years, and is expected to increase even further in the future with the emergence of 6G mobile connectivity technology and have become critical tool in generating massive amounts of data, which when processed and analyzed have enabled cities to improve on areas like branding through place identity. For instance, there are now cities like Amsterdam, New York and others relying on the power of social media to market, create and advocate for their place identity. Other cities, like Amsterdam, have been using the power digital solution to enhance its position as the ultimate world cycling city.

The advantage of this practice is that cities are able to draw not only tourism traffic, but also local ones, which ultimately help improve their economic status through increased job opportunity and increased revenue flow, while capitalizing on resiliency outcomes. The place identity concept also allows cities to create new or capitalize on their inherent and unique heritage like culture, art, education, health and others; thus, allowing them to remain competitive, while retaining their uniqueness and community heritage -which is often an issue in rapidly gentrified cities. Further to this, Parente notes that when place identity is well crafted, deployed and achieved, cities are able to attract new business opportunities, and in the long-turn manage to contribute toward improving the quality of lives of its residents, while reverberating positively on adjacent urban areas. The "15-Minute City" concept, on its right supports such drives and advocates for the promotion of place identity through the concept of chrono-urbanism, allow those seeking alternative lifestyles -away from the automobile dependent cities, to access almost urban services and amenities within walkable and biking distance.

Regarding the pursuance of social welfare, one area that is key is ensuring that residents are able to enjoy health lifestyles. Such are enhanced through the ability to walk or cycle to different parts of the city within the prescribed 15 min. Besides walking, we support that the substantial amount of time saved for not being in traffic could be directed towards engaging in other health -both physical and mental, enhancing activities. On this front, modern technologies have become accessible and practical, especially due to availability of wearables (including those gathering biodata) and urban sensors and data collection techniques that not only allow people to keep tab of their health, but also provide large datasets that could be used to improve the liveability of the area. Finally, the land use approach of this concept gains from the use of technologies as it can optimally ensure that cities have sufficient land reserves for playgrounds, recreation areas and foot walk and bicycle lanes to achieve the desired dimensions and for maintaining an optimal density that would be supported by available resources.. On this front, the use of Blockchain, smart contracts and fiscal instruments can be tallied to urban development rights and transaction costs to ensure both efficiency and performance while catering for long term policy agendas.

INTRODUCING THE 15-MINUTE CITY CONCEPT

The 15-Minute City rides on the concept of "chrono-urbanism", which outlines that the quality of urban life is inversely proportional to the amount of time invested in transportation, more so through the use of automobiles. This concept originated from the first author, Carlos Moreno, who advocates for an urban set-up where locals are able to access all of their basic essentials at distances that would not take them more than 15 min by foot or by bicycle. Interestingly, the concept is observed to emanate from his tenets developed in his version of a "living city", outlining how it is necessary to "repair" urban and social fragments, largely fueled by modernist approaches. For the present "15-minute" concept, Moreno supports that residents will be able to enjoy a higher quality of life where they will be able to effectively fulfil six essential urban social functions to sustain a decent urban life. Those include (a) living, (b) working, (c) commerce, (d) healthcare, (e) education and (f) entertainment.

To attain those functions, the urban built landscape needs to be restructured to ensure that it complies with components such as proximity, diversity, density and ubiquity which Moreno opines to be major in the pursuit of cities offering an urban life that could be categorized as being of high value. A case in point where this concept has been tried and tested is in Paris, where Anne Hidalgo, through the program "Paris en Commun", turned towards the "15-Minute City" concept, imagined by Carlos Moreno in 2016, in a bid to win a second mandate as Mayor, which she successfully did in June 2020.

This is quite a remarkable achievement, noting that citizens of a pandemic-stricken global city opted for a program highlighting dimensions of resilience and livability over pure economic growth.

Following its success there, this concept has now been seen to be replicated in different cities across the globe and aroused the interest of international organizations, which, amongst others, included the C40 Cities, the World Health

Organisation (WHO), UN-Habitat and the Organisation for Economic Co-operation and Development (OECD) in order to improve the quality of life in the context of COVID-19 and beyond.

While striving to satisfy those components, the six essential social functions are bound to be achieved, with the urban social fabric becoming even more closely knitted and with residents made to interact and participate more in activities that ultimately strengthen their social bonds, building character and trust, which ultimately leads to the building of more healthy urban landscapes.

In the current era where the world is experiencing the fourth industrial revolution characterized by ubiquitous deployment of Information Communication Technology (ICT), the 15-Minute City concept is seen as being timely. This is true since ICT is now seen to offer an array of solutions to numerous urban challenges, especially through the Smart City concept. With these technologies, it has been shown that cities could enhance and improve on service delivery as well as promote increasing sustainable practices. The prompt delivery of services is also at the core of the 15-Minute City concept with the ultimate objective of ensuring that the maximum time is available for urban dwellers to accomplish the aforementioned basic social functions. In particular, the time saved from mobility is argued to be critical as urban dwellers do not only waste time on the roads, but they also incur high costs during commuting as well as degrade their life.

To put this into perspective, in America, it is estimated that on average, traffic delays leads to a consumption of USD 1010 per year per person, which cumulatively translates to over USD 166 billion per year nationally. On the same note, an average national tally of 54 h a year are wasted by every average U.S. driver on the road due to traffic. Saving on commuting time also saves the urban environment through the reduced emissions from both vehicles and power plants when fuels are extracted and processed.

In addition to reducing time, the transformation that is to take place to ensure that all basic amenities are accessible within reach must ensure that the available urban space is optimized, hence promoting novel land use, which is also critical in reconciling human livability pursuits aligning with the urban ecology.

In respect to the above backdrop drawn from Moreno's 15-Minute City concept, Figure 1 showcases the proposed "modified 15-Minute City" framework, depicting the four identified dimensions that could be incorporated with the already existing one proposed by Moreno. These are (a) Density, (b) Proximity, (c) Diversity and (d) Digitalization.

The four dimensions were identified after observing the challenges that different cities across the globe endured during the height of widespread cases of COVID-19 and the subsequent health measures and protocols that ensued aimed at mitigating the spread. As noted above, urban residents endured uncountable challenges, especially in accessing basic essentials which, in most cities, have been sparsely distributed, whereas if those cities had prioritized the four dimensions identified here within a comprehensive and flexible framework, we believe that numerous issues could have been avoided.

Figure 1. The 15-Minute City framework

For instance, a neighborhood planned such that all basic facilities could be within reach in 15 min or less and that has public spaces ri allow people to maintain decent health standards and practices such as exercising through walking while reducing social contact would be relevant in the time of the present pandemic. Notwithstanding this, going forward, as urban areas embrace the "new normal", the proposed planning model that emphasizes the four components offers a realistic alternative.

Density

Density is a crucial dimension of the city and its built environment, and this has been noted by numerous researchers as having a direct link to travel and diversity. In conventional planning, density is viewed in terms of ultra-high-density buildings, but in the 15-Min City concept, density is viewed in terms of people per kilometer square. That is, in planning for a city that is sustainable, it is supported that it is paramount to consider the optimal number of people that a given area can comfortably sustain in terms of urban service delivery and resource consumption. In earlier planning models, where the emphasis was on creating ultra-high-rise buildings and offices, challenges arose, including the increased overconsumption of resources and over-reliance on fossil fuel energy to power buildings, ultimately leading to an increase in the number of automobiles inspired by centralized planning ideologies. As such, the emphasis here is on the optimal density that ultimately allows sustainability pursuits to be achieved on the economic, social and environmental frontiers.

Salingaros highlights that with an optimal density, it is possible to effectively plan the available space such that all the essentials could be accessible to residents without the need for time- and energy-consuming automobiles. Furthermore, he opined that an optimal density allows for the creation of locally based solutions in areas such as energy generation, food supply and multiple use of available spaces, where, for example, school playgrounds could be used as parks. In the current urban dispensation, this dimension allows for the creation of public infrastructures such as bicycle lanes and walkable pathways that minimize the requirement for automobiles, hence promoting the achievement of the social functions imagined by Moreno. Our view on this issue aligns with that of Sisson, who notes that this dimension also allows for equity in addressing the specific needs of different areas, including communities of varying economic status, while particularly favoring the disadvantaged, especially through the equitable distribution of civic services and in the deployment of services at a reduced cost for cities, rendering higher value for both investors and governments. Through this lens, the stand on this perspective paper is that density is a key component favoring the social sustainability dimension of cities, and when integrated with the other three dimensions (proximity, diversity and digitalization), the "15-Minute City" model would not only help in improving service delivery and improve livability but also benefit from the advanced technological advancements.

Proximity

This section promotes the dimension of proximity, which is viewed to be both temporal and spatial. That is, within the 15-min quickly accessible radial nodes, residents in a given neighborhood can readily access basic services. This dimension is critical not only in helping cities reduce the amount of time lost in commuting but also in reducing the environmental and economic impacts of such activity. This also helps in determining social indicators impacting urban residents, especially in a bid to promote social interactions—as advocated for by urban theorists such as Jane Jacobs, Christopher Alexander, Andres Duany and Nikos Salingaros.

On this, Moreno, advances that this dimension is critical as it allow residents to transition from residential areas, work, commercial areas, education centers, health facilities and other basic institutions in a reduced timespan. Ultimately, this closeness of essential services allows residents to enjoy better service provision both in commercial and public establishments, as the planning model allows for multimodal use of basic infrastructures.

For instance, in Paris, where Mayor Hidalgo has embraced this new concept, it is becoming a norm for school playgrounds to be transformed into parks that the public are free to access after school hours. With this and other examples, the pursuance of the proximity dimension advanced in this paper is, thus, critical, as it allows local residents to maximize their local public spaces, green spaces and other public infrastructures. It also allows them to maximize the exploitation of their available resources such as heritage for social, economic and environmental benefits. This premise is important as it allows to envision a city through a different perspective, more tailored to the human scale and adaptable to the use of its residents.

In essence, this concept of proximity as advanced in this paper links back to the core the philosophy of chrono-urbanism, emerging from the thematic of "chronotopy", standing as a coupling of spatial and temporal dimensions. Through this lens, the city can be viewed from a distance-based indicator, leading to a view of time-based urbanism. However, as is demonstrated in this paper, it is not sufficient on its own to actualize this node, as it needs to be connected to the other dimensions to achieve the desired urban transformation.

Diversity

Diversity in the context of the above frame and in the advancement of the 15-Minute City concept is twofold: (i) the need for mixed use neighborhoods which are primary in providing a healthy mix of residential, commercial and entertainment components and (ii) diversity in culture and people.

Having mixed-use neighborhoods is key in sustaining economically vibrant urban fabrics, ensuring sufficient housing for all the urban residents, promoting inclusivity and promoting sustainable practices. In the pursuit of a 15-Minute City model, the adoption of mixed-use neighborhoods is paramount in ensuring that an optimal density and proximity of essential amenities are achieved, while also providing for development of walkable streets and bicycle lanes. This approach ensures that residents can benefit from basic essentials within their residential areas, thus reducing the need for them to travel to access them. This dimension would also aid in ensuring that the sanctity of available public spaces is preserved and upheld and, where possible, that opportunities for creating more public areas are savored, as was advocated for by Whyte. This dimension aligns with that of urban density and proximity, and it is the position of this paper that emphasizing its observance in the proposed "15-Minute City" model would help enrich the experience of urban dwellers.

Brookfield expresses that embracing this form of diversity in a neighborhood provides city governments with opportunities to concentrate on improving their service delivery to locals and, in turn, promotes the livability status, maintains property value and enhances community participation and interaction. This is particularly important in promoting a city environment that accommodates different cultures and people, thus promoting social cohesion and helping in creating more social capital. Having a multicultural dimension in a city has positive bearings on the economic front, as locals would also enjoy a wide variety of products such as cultural products and cultural heritage. This would also render an attractive urban landscape for visitors, thus promoting tourism and other related businesses which are essential in creating new business and promoting economic vibrancy, leading to the creation of increased employment opportunities.

To maximize those benefits, this dimension requires deployment at varying scales in a city. Not only in an urban scale across a 15-min radius but beyond more minute scales—at building levels.

Digitalization

This dimension is very relevant to the modified 15-Minute City concept, especially in ensuring the actualization of the three other dimensions. In particular, this dimension aligns closely with the Smart City concept from which the 15-Minute City concept can be argued to have, in part, drawn its inspiration. For instance, within the Smart City concept, factors such as inclusivity, resident participation and real-time delivery of services are encouraged through varying platforms—including digital—just as in the case in Moreno's proposed concept. The achievement of those factors in the Smart City concept is shouldered on the effective deployment of different technologies, and this would have similar far-reaching impacts in the 15-Minute City. For instance, through digital tools and solutions, it would be possible to ensure that biking experiences are enhanced by emphasizing solutions such as bike sharing and the deployment of sensors to ensure the safety and security of cyclists.

Regarding proximity dimensions, digitalization has been effective, as highlighted through the Smart City concept, where services such as online shopping, cashless transactions and virtual communications and interactions amongst others are implemented and promoted. The similar availability of such services within 15-min cities would reduce the need for commuting as some services could be delivered within the comfort of homes or offices. Coupling these with technologies such as smart contracts through Blockchain technology would help to ease security concerns, especially in regard to virtual payments. Digitalization, especially during this period of the COVID-19 pandemic, has made it possible for people to work from home and communicate virtually, and this also been instrumental in reducing social contact and reducing the need to travel from home to offices and other work places. This dimension of proximity-based planning is already emerging from the pandemic and will be expected to stay in the post-pandemic urban jargon.

Additionally, the deployment of digital solutions would greatly help in even exceeding the expectations of the 15-min city, in particular in ensuring that cities are more resilient in areas such as climate change through declining emissions, linked to a reduction in automobile use. This would also be true where the adoption of mixed energy options is embraced, which help in reduction in over-reliance on non-renewable energy sources. Further, digitalization is crucial in facilitating optimal consumption of resources. Digitalization is also becoming very powerful in job creation through innovations such as iBike that allow people to share bikes in cities, hence encouraging them to avoid the use of personal cars. All those are expected to benefit immensely from the advent of digitalization, which has been seen to be a transversal element, cutting through fields, and as such, it is expected that it would also be essential in ensuring the successful deployment of the other three identified dimensions.

In pursuit of transforming urban areas to improve their livability status and resilience, promote sustainability and promote their social and economic aspects, many planning models were proposed over the years and have evolved as a result of their application in varying contexts and geographies. The most recent one is the Smart City concept, which, though it has been progressive in addressing most of the aforementioned issues, has had its shortcomings. From a theoretical perspective, it has the potential to help cities achieve all those objectives, but this potential has been overshadowed by enterprising orientations of ICT technology providers who are inspired solely by the lucrative profit margins that this model presents (valued at USD 410.8 billion in 2020 and is expected to rise to over USD 820.7 billion by 2025). As a result of this sole pursuit, based on economic agendas instead of solving issues such as non-inclusiveness and social inequalities, the Smart City concept, in some instances, has been found to exacerbate them. For instance, in regard to issues such as cost of living, Pandey notes that urban areas that have transformed and marketed as being "smart" are too expensive for most urban residents to afford. The author explains that this phenomenon is influenced by the emphasis on smart technology rather than the social and economic aspect that such technologies are supposed to address, thus deviating from the dictates of the SDGs, especially SDG 11.

As noted by Adkins, et al., the concept of the Smart City has acted as a catalyst for the growing inequality in the housing sector, where the real-estate sector is seen to be growing at a rate not proportional to the revenue growth of residents. The growth of the real-estate sector, which is majorly controlled by high-income earners, in turn, has been the fore in advancing the issue of inaccessibility in cities, as it also follows the grid planning systems that have been adopted in many cities. A similar argument is advanced by Gurstein and Hutton, where they note that modern urban planning concepts need to be restudied as they have led to unsustainable urban trends, including urban sprawl. The authors further advocate for planning concepts that would reorient urban development into paths that support both environmental and equitable paths, and the present authors believe that the proposed 15-Minute City concept could be among the fitting models.

Those planning weaknesses of most urban planning models, including the Smart City concept, were exposed when the pandemic emerged, necessitating the introduction of stringent health protocols and the re-alignment of economic agendas in view of the increasing socio-economic issues brought about by the pandemic. With the need to observe lockdown measures, social distancing, curfews and other health protocols, it becomes evident that the need for proximity of most basic amenities in most cities has been wanting. Thereby, most urban residents have had to put up with myriads of both economic and social challenges due to inaccessibility of most basic essentials despite availability of advanced soft Smart City technologies.

With those challenges, most cities, as currently planned and constituted, need to be re-structured to ensure that in the future, basic services are available at accessible distances, that human interaction would not be disrupted and that cycling and walking would be facilitated to ensure that such healthy protocols would be sustainably observed. Another emerging issue about the Smart Cities model, apparent during the COVID-19 lockdowns, is that most soft technologies support an increasing amount of urban data and their analysis, where this has not, however, directly led to more social interactions, which are now understood as an important planning aspect, especially to help residents weather down both the social and psychological challenges prompted by the COVID-19 pandemic. In light of this, some aspects advocated for in the "15-Minute City" concept gained traction during the pandemic, despite this model having been proposed as early as 2016. The selling point of this concept is its emphasis on proximity-based planning, where an urban neighborhood is planned to accommodate an optimal density that would have access to basic essential services within a 15-min walking or cycling distance. The proponent of this concept envisions that within that a 15-min radius, residents will manage to experience a higher quality of life as they will be required to travel less to access basic facilities such as public spaces, with increased time and opportunities to interact with other members of the community and accomplish other social functions, which are increasingly important but which have been lacking as a core function of contemporary urban planning models.

The emphasis on accessibility and proximity advanced in the 15-min concept, especially through walking or via cycling, is paramount, since this mode (micro-mobility) has been showcased as having numerous benefits on social, economic and environmental scales. For instance, cities benefit from reduced congestion, reduced pollution (noise, emissions and others) and benefit from beautiful increased green spaces, and well-thought and ordered structures. On the same note, urban residents derive health and economic benefits. They also benefit from increased time and opportunities to exercise and gain social interactions. On the economic front, this model has the potential to unlock numerous potential positive outcomes such as employment, new innovations, creation of urban unique (identity) brands as well as helping to reduce overheads incurred from fuel costs, road maintenance, pollution and other associated costs.

It is noteworthy that while the concept of "chrono-urbanism" may seem arbitrary for some—e.g., why 15 min and not 17 min?—this concept is not rigid in nature and is proposed with the intent to be tailored to individual cities based on both their morphology and specific needs and characteristics. For instance, on this, it is worth noting within 15-min radii, a cyclist would cover a substantial distance compared to residents who may opt to walk. Therefore, the proximity dimension for cyclists would be defined differently from that of those on foot. Indeed, on this matter, there had been other concepts such as 20-min cities and 30-min cities, but the bottom line in all of those is the need to underline that proximity-based planning is key in sustaining quality of life and in providing for the basic urban functions. On this subject, the current concept by Moreno is based on the understanding that within the 15-min radius, it is possible for a city to incorporate all the basic essentials, including walking and bicycle lanes, while also incorporating digital solutions to enhance the residents' experience and improve livability status. This thus enables works that support the spatiotemporal dimension of cities to both understand and enhance the quality of life of urban dwellers

As this concept continues to gain acceptability and traction in different global cities, further research is now warranted to showcase how the idea and its elements can be replicated in cities within the global south and those that may be financially constrained to undertake the extensive urban regeneration exercise that this kind of planning model demands. The research gains in importance, noting that some cities can transform and renovate existing infrastructures (without major restructuring) to benefit from bicycle lanes, parks and greener spaces to match the proposed model and to align with the vision of urbanists such as Whyte, Alexander, Salingaros and Jacobs. The pandemic will see long-term changes in cities, prompt the emergence of new socio-economic structures and force new urban planning models that need to adapt to such trends to ensure a safe and sustainable urban future.

Finally, it is observed that the 15-min city is in line with concepts promoting dimensions of proximity, emphasizing walkability and of social interactions within cities. Alexander, Gehl and Whyte seem to support this point through their treatises of human-scale urban design. However, even though the concept has gained a rapid adoption by global cities, we support that the operationalization of the 15-min city will require additional studies in order to scrutinize the respective importance and the connections of density, proximity, diversity and digitalization. To do so, it will be necessary to mobilize up-to-date research methods and to discuss its ability to complement classical modelization approaches. One therefore has to consider that beyond its attractiveness for Mayors and institutional organizations such as C40, the 15-min city is opening the way to a research program.

CHAPTER 3
LIFECYCLE

The end of a building's physical life happens when its technical condition is so poor that renovation is not financially feasible, or when its geometry does not support easy repurposing.

At this phase of the digital building lifecycle, a digital handover is created as a live archive. This snapshot in time captures the building's complete history, technical information, and profile and retains them as an invaluable postmortem sample to learn from when renovating similar buildings or creating new ones.

Long after physical buildings are gone, for example, machine learning algorithms and AI can be used to process and compare data about existing or planned buildings against stored outcomes from previously built projects.

Designers, contractors, or manufacturers can test expected ideas or outcomes, or explain why certain issues occurred and how they might be avoided in a current project. They can more quickly discover how buildings were made in the past in order to replicate successes or drive greater sustainability.

A DIGITAL PHASE IN PROGRESS

Digital handoff is an emerging step in the digital building lifecycle and requires some industry-level discussions on how to make it operationally possible. There is currently a lot of repetition in the built environment, and the industry is saturated with different standards that are not yet integrated.

Figuring out how to activate this important phase of the digital building lifecycle can free our industry of outdated processes that treat each project as something completely new, operate only in linear terms, and repeat mistakes of the past.

Our building lifecycles can become continuous and cumulative, enabling future designers and builders to launch new projects, innovate solutions, and transform built environments from all that has come before.

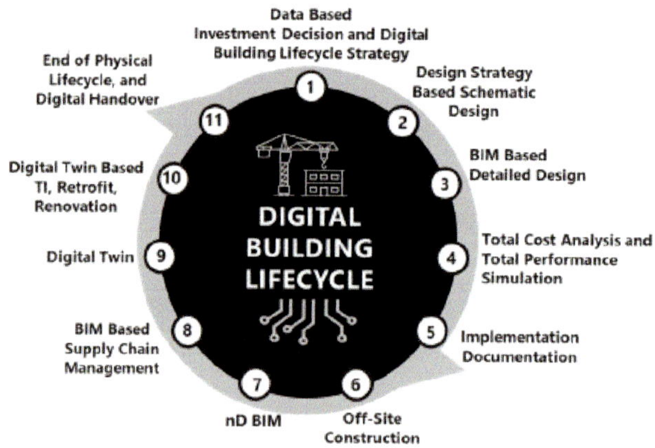

Fig. 1. Digital Building Lifecycle

LIFE-CYCLE OF A BUILDING

If our goal, as designers and developers of a sustainable built environment, is to design waste out of the building's life-cycle, it is important to understand the environmental impact of the different life-cycle stages. You can make more impact if you do not just think about the use stage of a building, but also about the production, construction and end-of-service stage during the design of a building. It is also crucial to understand which stakeholders are involved per stage because multidisciplinary collaboration is key within the circular economy.

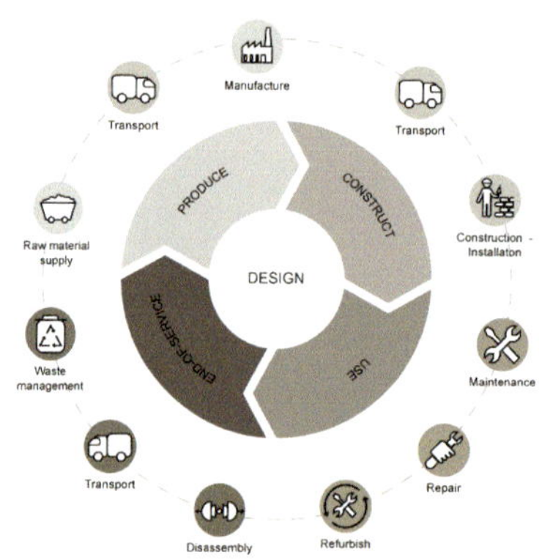

Fig. 2. Life-Cycle of a Building

We divide a building's life-cycle into four different stages: production stage, construction stage, use stage, and end-of-life stage. In the end-of-service the materials can become an environmental burden or have positive impact by feeding them back to the production stage, reducing the extraction of raw materials.

During the production stage, raw materials are extracted, transported and converted into construction materials. The extraction of natural resources has a great impact on the availability of non-renewable resource and might lead to their depletion. Besides, a large amount of water and energy are related to this process. The manufacturing of construction products also requires a large amount of energy as it has been estimated that the construction materials production industry is one of the biggest energy consumer at a global level. Moreover, energy consumption leads to the release of air pollutants.

The construction stage is when the building takes its form. The construction stage of a building requires a considerable amount of energy and materials, and involves a high number of stakeholders.

The use stage is usually considered as the longest in the building life-cycle. This stage includes activities such as maintenance, and it is characterized by circular strategies such as reuse, repair and refurbish to extend the life span of the building. The use stage leads to environmental impact caused by the users and the physical characteristics of the building, i.e. energy and water use and waste generation.

The end-of-service stage in the context of the Circular Economy would mean that the buildings can be safely disassembled into different components, hence they can be reused, repaired, remanufactured or recycled. Since we live in a Linear Economy, the end-of-life of buildings is the demolition process. Demolition creates huge quantities of bulk waste which might be reused, recycled or simply landfilled. The environmental impact from this phase is related to the release of green gas emissions from machinery and transportation, as well as emissions related to landfill disposal.

ENERGY RENOVATION AS A KEY ELEMENT IN URBAN REGENERATION

In June 2018, ROCKWOOL and the Building Performance Institute Europe, BPIE, published a report with a selection of successful urban regeneration projects and an analysis of the key elements in achieving successful urban regeneration.

Many large cities host run-down or challenging neighbourhoods, often characterised by multiple social problems, inadequate living conditions and bad reputations. These problems often reinforce each other in a vicious circle, causing a negative trend for the area (Fig. 3).

Poor houses ® energy poverty ® social problems ® vandalism ® bad reputation ® lack of investments ® poor houses.

The challenge to revitalise these areas is amplified by the fact that these neighbourhoods are often physically and socially detached from the rest of the city.

The buildings forming these areas are often 40–50 years old and were built at a time characterised by different societal ideals. During the 1960s and 1970s, there was a huge need for dwellings—resulting in most constructions being built quickly and on the idea of equality. Time has changed and so has our demand from the built environment. Most of these buildings are now in dire need of an upgrade to lower their energy consumption and improve their thermal comfort and the health conditions of the people living in them, as well as to upgrade the whole neighbourhood that they are part of through the improvement of their aesthetic quality and that of the urban spaces.

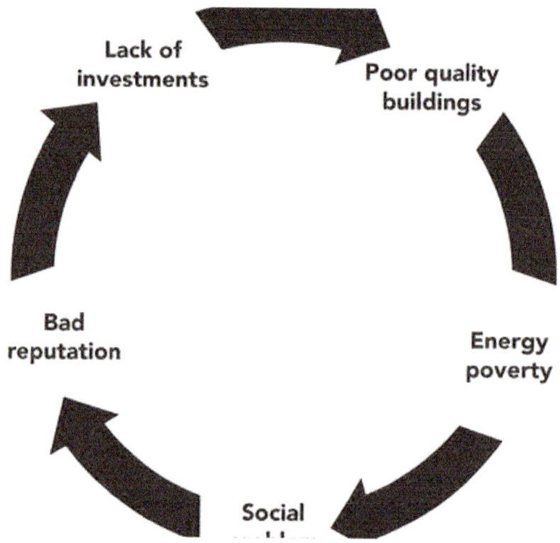

Fig. 3.The vicious circle of poor housing in challenging neighbourhoods.

Urban regeneration deserves additional attention as cities need to be transformed into living spaces and working spaces in a sustainable manner. Often, challenged urban areas especially need attention in creating resilient buildings as the combination of high-rise buildings and dense urban areas poses a higher risk for the built environment in case of fire or natural disasters. An upgraded building stock will also enable and induce the citizens to live a more sustainable lifestyle.

Many case projects around the world have shown that a comprehensive strategy comprising both physical and social initiatives can transform whole areas into attractive and liveable spaces as well as turn around the negative trend experienced in these neighbourhoods. Upscaling energy renovation of run-down or challenged urban areas is a key element, which can, if combined with other initiatives, not only improve the quality of life and enable citizens to live a sustainable lifestyle but also provide an opportunity to reduce social and health problems in society. Upscaling energy renovation should at the same time focus on creating new business models out of the challenges—focusing on the need for managing resources in the urban environment in a better way.

Researchers describe a selection of successful urban regeneration projects in their study. The key elements in achieving successful urban regeneration summarised later are the result of a review of successful cases combined with an interactive dialogue with leading experts in the field facilitated in 2018.

The key learnings identified are especially:

- A strong long-term commitment of and an early engagement from the public authorities is necessary.
- A well-adjusted combination of social and physical measures needs to be considered—which also requires an early collaboration between multiple actors with different expertise.
- The involvement and empowerment of the people living in the area are crucial. Engaging residents in the renovation process, e.g. via 'social contracts', increases the support among residents and can help to give people a new start.

Key success factors which were identified for the building renovation projects were:

- A detailed assessment of renovation alternatives found that a deep renovation would be the cheapest option over a 30-year period.
- Improving the aesthetic quality of the areas combined with the use of long-lasting materials is cheaper in the long run and can attract new investments and new residents to the areas. Doing so can also be a tool to reduce crime and improve quality of life in the area as neighbourhoods that are in a poor state are perceived to be uncared for—resulting in the likelihood of crime going unchallenged.
- Building owners see an increased value of their properties while reducing operational and maintenance costs.

Key elements in achieving successful urban regeneration

One of the key learnings from the many existing urban regeneration projects is that people in general like their homes, want to be informed and involved in large-scale renovation projects and prefer to be given the opportunity to take individual choices.

A deeper dive into the findings provides the following insights.

Urban regeneration is not the same as gentrification. It is, therefore, important to make sure the residents are involved and empowered in the regeneration process. While many of the run-down or challenged areas would benefit from hosting a more mixed population, the means to achieve this cannot be to squeeze out certain groups. Extensive energy renovation will actually improve residents' ability to pay rent through lower energy bills. A successful urban regeneration process can therefore lift a whole neighbourhood including the people in it.

A successful urban regeneration strategy cannot be fulfilled without really knowing the needs of the residents. A bottom-up approach where residents are involved can uncover new innovative solutions while giving the residents a sense of ownership of the process, thus increasing their willingness to participate. In other words, an urban regeneration strategy must build on the existing culture rather than building something new. Large-scale renovation projects can be tailored to the existing culture, e.g. by developing a catalogue of individual measures.

Look beyond the near-term period. A long-term plan for the neighbourhood will increase the chances of a successful regeneration process. The plan should incorporate the multiple benefits that a physical and social transformation will entail such as a higher living standard and improved health, the alleviation of energy poverty through lower energy bills, an increase in housing prices, the reduction of maintenance cost and the mitigation of the area's climate impact.

Furthermore, the multiple benefits of deep renovation should be valued and monetized. Research from Eurofound shows that inadequate housing is linked with numerous societal costs ranging from health care expenses, higher policing and emergency costs to poor academic performance. Public authorities ought to incorporate these costs and benefits in their cost–benefit analysis.

The physical transformation of a home and a neighbourhood can have a positive impact on people's perception and behaviour. A balanced urban regeneration process needs to combine both physical and social measures in a harmonised manner. Physical and social measures are not two separate entities but are highly interlinked, as a new home is commonly associated to new beginnings.

The physical transformation of buildings should be long-lasting. It is evident that deep renovations using high-quality materials, combining the energy upgrade with an architectural upgrade, are the cheapest solution in the long run. Deep renovation guarantees a very low energy demand for the next 30 to 50 years. In addition, maintenance and administration costs are reduced as well. Past experience has shown that renovation projects that were quickly done using cheap materials had a short lifespan—thus requiring renovation again. In contrast, high-quality and durable materials guarantee that renovated buildings continue to look nice for many decades.

In all the cases investigated, the cities/municipalities have played a key role. The authorities have been successful in driving transformation through public-private partnerships, in which they safeguard the social aspects. Another key success factor is the breaking down of silos within the public governance, for example, by setting up a task group with various departments (energy, climate, social, budget, etc.) and involving solution providers early in the process. Scaling up deep renovation requires more collaboration between various stakeholders, assembling different skills and expertise.

Good planning can use the 'economies of scale' effect when renovating multiple identical buildings. The buildings built during the 1960s and 1970s are especially suited to a more industrialised renovation approach, which could reduce the time spent on site and cost. New business models (one-stop shops) and technologies should be supported and further explored.

Moving forward

It is evident that extensive experience exists from individual case studies and within individual stakeholders. The existing experience must be scaled up—following the success factors from existing urban regeneration projects. Cities should be encouraged to develop their own urban regeneration path and to take responsibility for facilitating an early engagement with stakeholders covering all needed competences in a project.

SUSTAINABLE BUILDINGS AS CONTRIBUTORS TO THE CIRCULAR ECONOMY IN URBAN REGENERATION

The benefits of energy renovation in urban regeneration are not limited to energy savings alone. It is broadly recognised to also benefit people and the environment. Good indoor climate, thermal comfort, acoustic performance and daylight, for example, improve the health and wellbeing of inhabitants and positively affect the productivity in schools, hospitals and of workers in general. Taking a life-cycle approach when considering the materials used in buildings can reduce the environmental impact of the buildings in a city. Moreover, valuing the potential for today and the future creates better economic value. Not surprisingly, there is clear trend towards a more sustainable design and renovation of buildings.

Urban centres are particularly impacted by waste from construction and demolition activities (CDW). The Ellen MacArthur Foundation has identified the built environment to be one of the key sectors of structural waste. For example, in Europe, the average office is used only 35–50% of the time, even during working hours. The UN-IRP study called 'The Weight of Cities' shows that the global 'domestic material consumption' (DMC) of raw materials (including sand, gravel, iron ore, coal and wood) is likely to be in the range of 8–17 tonnes per capita per year at 2050, assuming material use per capita will stabilise in developing countries at lower levels than today's developed countries.

However, a DMC range of 6–8 tonnes per capita per year has been proposed as an indicative target for sustainable resource consumption. In 2050, urban mining should be the main source of building materials in cities, with implications for storage, logistics and costs, while citizens will increasingly require a non-toxic environment.

Cities are well placed to tackle the resource and waste issues in the construction sector, with their high concentration of resources, capital, data, and talent over a small geographic territory. Closing the loop in the construction sector and improving design would lead to major benefits in sustainability and in quality of life. This could further support urban policymakers in achieving their objectives when it comes to carbon emissions, mobility, indoor air quality and working towards a non-toxic environment. This transition towards a sustainable, low-carbon and resource-efficient economy—called a 'circular economy'—is considered vital to future-proofing cities and improving quality of life for citizens. Our world economy is considered to be only 9.1% circular at the moment.

The Circle Economy states that 'Closing the circularity gap serves the higher objective of preventing further and accelerated environmental degradation and social inequality'—especially since housing and infrastructure needs represent the largest resource footprint. They believe that a circular economy approach enables cities to take practical steps to help reduce emissions, create new jobs and strengthen industries and competitiveness, as well as enhance the health and wellbeing of its citizens. Sustainable buildings contribute to this transition.

Sustainable buildings

There is not one definition for 'sustainable buildings'. However, what all descriptions have in common is that a holistic approach is taken in the design, construction and demolition processes to minimise the buildings' impact on the environment, the occupants and the community and to maintain economic value. When it comes to the environmental aspect of sustainable buildings, the World Green Building Council describes a 'green' building as 'a building that, in its design, construction or operation, reduces or eliminates negative impacts, and can create positive impacts, on our climate and natural environment. Green buildings preserve precious natural resources and improve our quality of life'.

There are several features that can make a building 'green' and additional features to make them more sustainable. Environmental features pertain to resources like materials, water and land and energy and carbon, as well as pollution of air, water and soil. Health-related features pertain to indoor climate, acoustic performance and thermal comfort. Additional features could be related to risks like earthquakes and fires—addressing the resources that the building represents.

Several tools and sustainable building rating schemes are in place to assess the sustainability of buildings. Examples of European-wide used commercial rating schemes are BREEAM, DGNB, HQE and LEED, but local schemes exist too. The European Commission recently launched a beta version of a framework for sustainable buildings called Level(s), which describes the basic relevant features for the European housing stock. In the Netherlands and France, building legislation is in place to assess environmental building features, and in Germany, a similar approach is taken for the Green Public Procurement of buildings.

The key step to a sustainable design is to reduce the amount of energy needed to operate a building. Typically, 80% of the energy used in a building results from heating, cooling, hot water and lighting, while the remaining 20% is used for construction materials, transport and demolition. However, moving towards a more energy-efficient building stock will change this ratio over time—putting more focus on the materials and the 'beyond energy' features of sustainability. Environmental assessments of buildings are therefore based on a life-cycle approach using various environmental indicators.

Buildings as Material Banks

Considering that a so-called circular economy is vital to future-proofing cities and improving the quality of life of citizens, according to, organisations need to stimulate the development of a circular economy by developing concepts for cities to become more circular. The Ellen MacArthur Foundation describes the built environment in 'circular cities' as 'designed in a modular and flexible manner, sourcing healthy materials that improve the life quality of the residents, and minimise virgin material use.

It will be built using efficient construction techniques, and will be highly utilised thanks to shared, flexible and modular office spaces and housing. Components of buildings will be maintained and renewed when needed, while buildings will be used where possible to generate, rather than consume, power and food by facilitating closed loops of water, nutrients, materials, and energy, to mimic natural cycles'. This description shows several aspects of buildings in circular cities, centred around the basic assumption that buildings are '*material banks*'.

Today, buildings are often created with only one function in mind, like being a school, an office or a multi-family home. When societal needs or user preferences change, these monofunctional buildings usually become outdated or even obsolete—resulting in a high rate of building vacancy and premature demolition. Those buildings, no longer suitable for use, are considered as waste and are thus treated as such. By applying circularity principles both to the existing building stock and to new buildings, the concept of waste can be eliminated.

This principle is elaborated in the European project Buildings as Material Banks (BAMB). According to the BAMB vision, the term 'Buildings as Material Banks' refers to a materialised investment: "the building itself is considered as a materialised savings account for material resources, through which building materials, products and components are temporarily 'deposited' into a functional element or part of the building.

When socio-economic conditions are favourable, (a part of) the materials, products and components may be retrieved for another investment, that is: another building or another high-quality application. Seeing material resources as a temporary way of materialising investments opens the door to a wide range of circular business models, in which economic and environmental value is conserved and created through the reuse of materials, products, components and buildings, while (performance-based) services are provided to support the daily life of (end) users."

Three major systemic changes have been identified to support the BAMB vision:

- Change in design culture
- Change in value definition
- Change in collaboration across all actors

Both the existing building stock and the new buildings are interlinked in urban regeneration.

Resources from the existing building stock

The growing demand for and the scarcity of resources, less space for landfill and the challenges regarding the movement of resources through the city all create a push towards an approach of using the materials and products in existing buildings in the most optimal way. 'Urban mining' is mentioned in several sources as a necessary approach in circular cities. At the same time, it is identified that there are insufficient economic and regulatory incentives as well as a lack of trust in the quality and availability of secondary materials to create a market for recycled building materials. A lack of demand for reused or recycled products in combination with easy and cheap waste landfill and incineration hamper a sound business model throughout the construction chain.

The EU Construction and Demolition Waste Protocol lists seven features for successful implementation of recycling:

- Ban landfilling of construction and demolition waste.
- Implement mandatory pre-demolition waste audits (identify quality and quantity).
- Enforce traceability of waste to establish confidence (especially during transport).
- Urban planning (recycling facilities within city limits).
- Manage quality of recycled waste (same quality standards as virgin materials).
- Environmental management along the entire value chain.
- Create open markets through leading example of public procurement.

Below, possible measures are discussed to keep material resources from buildings in the loop, for both the end of the building's life and the production of new products

Measures at the end of the life of a building

Upon demolition, budget constraints do not allow for (often more labour-intensive) the deconstruction of separate material fractions. The limited space for collection of many fractions is a further barrier in cities—requiring a more sophisticated after-sorting. Toxic legacy materials require specific attention if products must be reused or recycled. In many countries, landfill or incineration are still easily accessible and are too cheap to incentivise economically feasible alternatives. Separation and recycling only work business-wise in countries or regions with strict landfill bans for recyclable materials—like in Denmark or the Netherlands.

This must be combined with (legal) requirements or incentives for buildings to be stripped before their frames can be demolished following an assessment (pre-demolition audit). This can create an entire service sector of SMEs with expertise in the removal of tiles, flooring, window frames, insulation material, sanitary items, lighting, etc. The Rotor in Flanders, Belgium, is an example of such a specialised company, as presented on the event from GLOBE-EU.

Cities can lead the way by discouraging landfill and incineration of construction and demolition waste and stimulating deconstruction, sorting and facilitate reuse platforms like Rotor. Local urban initiatives also contribute to social cohesion and job creation.

Measures to stimulate the use of secondary materials

Closing the route towards landfill and incineration must go together with boosting demand for reused products and secondary materials. The market for secondary raw materials is still immature, and several policy choices exist to boost this market. The policy toolkit report from the Ellen MacArthur Foundation explains it as follows: 'There are two broad, complementary policymaking strategies that can help accelerate the circular economy. The first is to focus on fixing market and regulatory failures.

The second is to actively stimulate market activity by, for example, setting targets, changing public procurement policy, creating collaboration platforms and providing financial or technical support to businesses. These approaches are complementary and policymakers can determine where to put the emphasis, taking inspiration from the most applicable aspects of both approaches'.

Within public procurement, cities have the power to set high thresholds for recycled content in their public tenders and encourage the use of reusable and recyclable materials and products. Other suggestions are fast-tracked approval of construction permits, easier access to capital and stimulating new ownership models of buildings and constructions. Creating platforms for collaboration and promoting value networks instead of value chains could also be a role for cities, as shown by the city of Brussels in the BAMB project and the Dutch city of Almere in their UpCycle City contest. A key takeaway from the Amsterdam Circular is to work closely together with the private sector and research institutes to find new business models in the context of existing strategies, such as green procurement.

Note that there is a growing belief that markets should be created without government interference like subsidies and levies on raw materials but much more by facilitating the emergence of business models that can work over time.

New buildings prepared for the future

New buildings can be requested to be designed and built in a 'circular' way. Better information about a building would help close the loop, and lower costs as worst-case scenarios in the future need no longer apply. Digital tools like BIM and cloud-based platforms of comprehensive data on buildings can already be used for today's buildings.

Reversible design and open building concept

Modularity and flexibility are mentioned by the Ellen MacArthur Foundation as key to circular buildings in cities. A Dutch Architect Manifesto refers to both adaptability/flexibility and easy-to-dismantle construction. Research on such reversible building designs is broadly available, as shown in the BAMB project and the EU-based project, although it is not a common practice yet. The Flanders region in Belgium has applied these principles in their policy for 'Building for change', a policy that anticipates on future changes in building use, as presented in.

An 'open building' concept is based on similar principles, thereby taking the perspective of ownership. This seems especially interesting in the urban environment, with many rental buildings and apartment blocks. This concept suggests that buildings need to be looked at from the perspective of the owner (long-term client) and the tenant (short-term client), respectively. The diverging interests of these two owners make it important to treat them separately (two-step housing). Investors in real estate are interested in the 'shell' of a building, not the interior outfit, which is considered more of a liability.

Cities in Japan are more advanced in the implementation of an open building concept which recognises that people need to adapt their living space according to changing circumstances. An entire industry sector has emerged around the remodelling of homes based on circularity principles: easy-to-remodel interiors, reuse of components and floor space tailored to family size.

Encouraging innovation by adding reversible design to the conditions for competitive bidding on public works contracts and the development of EU-wide indicators and standards for reversible building design would go a long way towards more sustainable building design and use.

Information throughout the chain

Information on the materials in a building and on the building itself over time facilitates a building to be a material bank in future. The digitalisation of society will support this.

Material passports are based on the principle that one should know what type of materials is applied in the building. Information on content should reveal potential future legacies of hazardous materials that could hamper recycling. Further information, e.g. on recyclability, is suggested. It is likely that building information modelling (BIM) will also play a role in the transfer of environmental product information through the chain, as it already fulfils this role for technical characteristics.

Whereas material passports are typically connected to products and their manufacturers, building passports could play a role in city policies. Building passports provide information on the whole building design, its maintenance, refurbishments, etc. Building passports are not well defined yet and are suggested for different purposes, including renovation and circular economy, but they could also be a helpful tool for risk evaluation in case of, e.g. a fire or natural disaster. An example of a building passport in the context of the circular economy is Madaster. It is Madaster's mission to eliminate waste by providing materials with an identity. The Madaster Platform facilitates the registration, organisation, storage and exchange of data of buildings over their lifespan, thereby becoming a public, online library of materials. The concept of building passports needs further development to be used in a consistent way on a large scale, but the principle clearly has a place when developing more circular buildings in an urban context.

What is a circular building?

It could be helpful in procurement and design processes if a definition of 'circular buildings' would exist. This is not the case, as the possibility of having a singular definition is being challenged, and the idea that 'circular buildings' are multidimensional is being considered.

Common themes in the descriptions of 'circular buildings' are energy efficiency, the materials applied in the building (their use of resources, reuse and recycling options and the waste they cause) and a design that enables long-term benefits from the building and its components. Such a 'circular building' is not similar to a 'sustainable building': sustainability is to be assessed, but a circular building puts more focus on design and saving resources while not compromising sustainability. For the environmental assessment, an LCA approach as described before is indispensable. LCA reveals any trade-offs of specific measures targeting specific issues, like resources or waste in the circular economy. In this regard, it is important to remember that the circular economy and circular buildings target a goal—usually of reducing resource consumption and waste—*but 'circularity' is not a goal in itself*! Not all circular concepts will result in environmental benefits in the end. It can be questioned, for example, if materials should be recycled into products that are not recyclable anymore—a pitfall for some innovative products made from waste.

RESILIENT BUILDINGS

Resilience

The term resilience has been around for centuries and has been used in engineering, ecology and psychology before coming to civil protection and disaster risk reduction.

Several definitions of the concept of resilience can be found in the literature, for example, in publications from Timmerman and Holling. The UN Office for Disaster Risk Reduction came with the following definition in 2009:

"The ability of a system, community or society exposed to hazards to resist, absorb, accommodate, adapt to, transform and recover from the effects of a hazard in a timely and efficient manner, including through the preservation and restoration of its essential basic structures and functions through risk management.".

Today, increased risk to hazards is manifested in cities. Climate change (flooding), earthquakes and terrorism, for example, pose new challenges to cities. Cities should be resilient to prevent social breakdown, physical collapse or economic deprivation. The Rockefeller Foundation report for the City Resilience Framework describes city resilience as 'the capacity of cities to function, so that the people living and working in cities – particularly the poor and vulnerable – survive and thrive no matter what stresses or shocks they encounter'. It is stated that risk assessments and measures to reduce specific foreseeable risks will continue to play an important role in urban planning. Buildings also have an important role to play in this area. In land-scarce urban areas, more and more people are working and making their homes in high-rise buildings, which impose a challenge on safety issues if not well taken care of. If an earthquake or a fire strikes in a dense city, the consequences can be very serious. Energy efficiency and sustainability should—and can—go hand in hand with more resilient buildings.

Resilient and sustainable buildings

In resilient cities, man-made infrastructure and buildings are well-conceived, well-constructed and safeguarded against known hazards. Building codes and standards should promote long-term robustness, flexibility to adapt in the future and safe failure mechanisms in the event of a shock.

It has been argued that design for resilience is to design for sustainability to reduce the environmental impacts and societal consequences of post-hazard repairs.

Conversely, design for sustainability is to design for resiliency to prevent that the unlikely extreme events may impact the urban communities. This study shows such an integrated approach for a holistic building design, considering safety, resiliency and sustainability, based on multiple conflicting criteria. The US building assessment scheme RELI has its own system of not only assessing all types of risks related to resilient buildings but also starting from the sustainable building perspective.

It should be realised that building codes and standards often provide minimum requirements, focusing on safe escape of people. Standard EN 16309 for assessing sustainable constructions starts from this principle, by valuing 'above codes' performance as being more sustainable. Accidental actions (earthquake, explosions, fire, traffic) are sustainable building aspects to assess. This shows once more the interlinkage of resilient and sustainable buildings.

The new EU Energy Performance of Buildings Directive (EPBD) from 2018 may lead to some improvements towards holistic building design, taking both energy performance and hazards into account. It requires for renovation projects to address fire safety and risks related to intense seismic activity affecting energy efficiency renovations and the lifetime of buildings, as well as the issue of healthy indoor climate conditions.

Fire-resilient buildings

The above section clarifies that fire resilience also has a connection to sustainable buildings. The fire resilience of a building is often described as the ability of a building to withstand the effects of a fire, or it is often linked to the building's fire resistance properties. However, fire resilience covers more than the technical characteristic of being fire-resistant—it considers how the environment, the community and economy adapt and recover from a fire.

Fire can cause disturbance to a city's ability to deliver its service to the community. Especially in densely populated urban areas, the built environment is a crucial part of the city's infrastructure. In January 2018, a fire in a hospital in South Korea killed 37 people and injured over 100. All patients had to be evacuated and the hospital was immediately closed. This meant that this hospital was no longer able to serve the community, and the remaining patients had to be relocated somewhere else. To improve resilience, it is recommended to include property protection of critical infrastructure, in addition to the current requirements for life safety in case of fire.

Also, efforts to improve the sustainability of buildings often focus on increasing energy efficiency and reducing the embodied carbon. However, a fire could reduce the overall sustainability of a building through the release of pollutants and the subsequent rebuild. Furthermore, fires have a range of less immediate and obvious adverse consequences on the natural environment. These include air contamination (which is likely to also include land and water contamination), contamination from water runoff containing toxic products and other environmental releases from burned materials.

As stated earlier, there is a disconnection between sustainable building approaches and building regulations. When designing a sustainable and energy-efficient building, architects usually focus on elements such as using solar energy and promoting air circulation. In parallel, fire safety requires the respect of building codes and other specific national requirements focusing on other elements such as safe escape routes and optimal operating conditions for fire fighters.

Moreover, sustainable building assessments are based on a life-cycle description of buildings, which is scenario-based—scenarios that do not plan for fire. However, statistics show the scenario of a fire occurring during the lifetime of a building is a realistic scenario altogether. In the USA, for example, an estimated 380,200 residential building fires were reported to fire departments each year between 2013 and 2015.

Therefore, methods that include quantitative risk assessment (QRA) for predicting fire spread to adjacent structures; life-cycle assessment (LCA) for estimating the environmental impact, the fire response and replacement of damaged materials; and cost–benefit analysis (CBA) for estimating the economic impact of the fire have been proposed to analyse optimal designs and environmental consequences. These methods could be one solution to include fire resilience in sustainable building assessment.

In summary, fire safety, sustainability and energy efficiency should be assessed together. A risk assessment should be performed both when building new and when renovating as fire does not only have an impact on the structure of a building but also has societal, environmental and economic impacts that we should try to mitigate.

CHAPTER

4 LIFECYCLES: THE FINE PRINT

Addressing increasingly pressing global challenges of climate change and other long-term impacts demands critical understanding of the sustainability (environmental, social, and economic) performance of current best practices of developing and managing the totality of urban infrastructure systems and accounting for how future technical and nontechnical changes influence the sustainability of cities. Dominant practices currently focus only on one aspect of the sustainability (e.g., environmental) or only one component of the city at a time (e.g., transportation) or only part of the life cycle of the infrastructure (e.g., embodied energy).

Decisions based on such partial information and limited knowledge should be replaced by more complete knowledge as a basis for decision analysis of the development and management of future infrastructure systems in cities to curb damaging and costly suboptimal developments.

A comprehensive account of relevant environmental, social, and economic aspects of our urban systems provides municipalities and high levels of government with the opportunity to look at the merits, synergies, and trade-offs associated with alternative development and operational decisions in creating sustainable cities.

What does a sustainable city look like? The literature offers as little on a widely accepted definition of sustainable city as it does on what sustainable development in general is. The difficulty of defining sustainable development has led to over 200 definitions.

Closer examination of the literature over the past 20 plus years shows that different issues are emphasized in the urban sustainability discourse depending on field of study and the problem at hand. Land-use system as a vehicle for achieving urban sustainability was not widely supported with evidence as mentioned in work published in 1997. When areas where improved environmental performance is needed in cities were identified, the emphasis was on how to integrate them with social, economic, and political goals of sustainable development. In a case study on the Scottish capital of Edinburgh, a work on weak and strong sustainability measures for cities differentiated between unsustainable cities, weak sustainable cities, and strong sustainable cities.

The strong sustainable cities model is close to what is now promoted as urban scale equivalent of circular economy at the majority of materials reused, repaired, and recycled, recognizing the limits to resources and assimilation capacities. The literature around sustainable cities reflects the compact city concepts of abating urban sprawl, increasing density, and achieving mixed uses while ensuring diversity and vitality, both socially and economically.

Urban form is an important element that has attracted significant research attention in the literature. Policies that focus on sustainability of cities should stay away from the quest for a long-term goal of creating a single definitive urban form that is sustainable as it does not exist. There are more avenues that lead to more sustainable forms. In their research that involved many British cities on the role of urban form, their conclusion was that "there are no simple answers or clear relationships between urban form and the dimensions of the sustainable city." Others have also explored how density and other urban forms relate to sustainability.

Two important aspects of urban sustainability are the issue of system boundary and the issue scale. Allocation of greenhouse gases (GHGs) from surface and air transportation spatially to cities that make up a commuter shed is demonstrated in. Factors such as surface vehicle miles traveled per capita (estimated to range from 13 to 129 km/capita/day for the Denver commuter shed) that affect the allocation are identified. Along the same line, geography, urban form, and the urban economy affect the GHG emissions of cities were highlighted elsewhere.

While liveability of cities is seen within a broader framework of urban sustainable development, the role of sufficiently incorporated urban natural areas in creating sustainable cities is also stressed. Sustainable cities are also presented in different studies as low-carbon cities (e.g.,). As part of achieving sustainable development of Chinese cities, low carbon goals of urban development are proposed by local governments. Governance level is considered a critical element in other parts of the world as well.

They provide the examples of "garden city, Chinese electric valley, and Sun City." Examining how the concepts of compact city, the zero-carbon eco-city, and the U-ecocity contradict or fall short of meeting sustainability principles, the use of the term "transition toward the sustainable city" was proposed as an accurate and more effective goal of our cities around the world aspire to be. Arguments for establishing guiding principles at different levels in creating sustainable cities are abundant. The literature is full of connecting the concept of sustainable cities to the triple bottom line of social, economic, and environmental standing of different aspects of urban areas. The need to convert our unsustainable cities by developing sustainable city models and rehabilitation plans that help us implement sustainability practices considering environmental and social performance factoring in economic elements is pointed out.

Earlier, the social, economic, and environmental dimensions of urban sustainability feature both complementarity and contradiction, underlining the importance of investigating alternatives using a holistic approach. Even much earlier, going beyond the triple of bottom line of social, economic, and environmental sustainability, the n-bottom line framework of performance attributes used in the research, development, and demonstration activities under the sustainable cities research theme at CSIRO, Australia, was presented.

Overall, there is a recognition that cities are better supported with improved sustainability-oriented policy-relevant information and knowledge on urban design, development, and management. There are potential multiple features that can represent what can be considered as part of the sustainable city vision.

From a life cycle perspective, a sustainable city is here defined as one whose residents and decision-makers account for a life cycle triple bottom line (social, economic, and environmental bottom lines) performance in prioritization and ranking of alternative decisions emphasizing the process. A sustainable city as a product is one that advances socially and economically with a nondeclining environmental quality over time.

Environmental sustainability that embodies environmental quality, or rather the avoidance of the degradation of it, is measured in the form of environmental impacts such as climate change, acidification, eutrophication, fossil fuel energy use, resource depletion, smog, ozone layer depletion, and different forms of toxicity. The level of impact should be capped within the city's allocated share based on what is required for achieving global net minimum sustainability threshold. Economic progress as a proxy to economic sustainability in a city in this case is measured as increase in GDP per capita together with very low or decrease in GINI coefficient implies.

The economic progress should be at the level where the city continues to make an operational profit that enables it to sustain its activities indefinitely. Social sustainability is represented by different aspects of social well-being such as inclusiveness, participation, employment opportunities, education successes, and health service coverage. It can be monitored by higher level of measured resident's satisfaction with sustaining their social well-being at the time of measurement and in the long run.

The rest of the chapter continues with a brief outline of challenges in cities, presentation of some life cycle opportunities, discussion of areas to be considered in achieving a life cycle sustainability of cities, and ending with the way forward.

DEFINING A SUSTAINABLE CITY

A consensual definition of sustainable cities does not exist, although most organizations share the same viewpoint. The Institute for Sustainable Communities (2017) published a concise proposal based on the Sustainable Communities Task Force Report (President's Council on Sustainable Development, 1998): "A sustainable community is one that is economically, environmentally, and socially healthy and resilient. It meets challenges through integrated solutions rather than through fragmented approaches that meet one of those goals at the expense of the others". This definition calls for an integration of the policies and strategies aimed at improving the sustainability of key urban issues. However, these are typically dealt with independently to solve specific problems, such as the access to public transport or waste management.

To identify the strategies proposed for each urban issue in the development of sustainable cities, we screened a set of fifteen reports and data from international organizations. Subsequently, we identified 28 potential improvement strategies that refer to 8 urban issues, namely buildings, energy, food, green spaces and landscape, mobility, urban planning, waste and water. In general, most strategies point to an increased process efficiency, the implementation of greener materials/designs and the integration of new technologies into the existing pool of alternatives.

Life cycle thinking (LCT) applied at the city scale

The feasibility of each strategy in environmental, economic and social terms is key to ensuring a sustainable urban performance. Quantitative studies can help to conduct this assessment, and LCT might be a suitable tool. According to the UNEP/SETAC Life Cycle Initiative LCT is "about going beyond the traditional focus on production site and manufacturing processes to include environmental, social and economic impacts of a product over its entire life cycle". This approach can be applied at different scales that range from single products to more complex systems, such as cities. Within this framework, four methods were developed to quantify the dimensions of sustainability. Life cycle assessment (LCA) (ISO 14040:2006) and life cycle costing (LCC) (ISO 15686-5:2008) quantify the environmental and economic impacts of a product system from raw material procurement to end of life, respectively. Social LCA (S-LCA) and life cycle sustainability assessment (LSCA) are newer methods, but their application is still incipient and needs standardization. LCT is the core of several policy action plans, including ecolabels and eco-innovation strategies.

Indeed, as stated in previous literature the development of a comprehensive methodology for the implementation of LCSA is still lacking, and issues regarding indicators, weighting and trade-offs between validity and applicability must be addressed. S-LCA needs further development, facing major challenges with regard to representation of results and data quality, among others.

However, LCT has been applied differently to the various issues involved in urban metabolism, leading to unbalanced data availability for evaluating sustainable strategies. Although the methodological limitations of LCT are well-known among practitioners, there is a need to identify the shortcomings of the LCT application to sustainability aspects related to cities, including both the topic coverage and methodological advancements.

LCT RESEARCH ON URBAN SUSTAINABILITY

This section presents the results of the review and discusses the state of the art for the following issues: buildings, energy, food, green spaces and landscape, mobility, urban planning, waste and water.

Buildings

The most evaluated strategy is the efficient use of energy, water and other resources in buildings (B1). Based on the nearly zero-energy building (NZEB) concept, LCA methodologies, and more specifically the life cycle energy assessment (LCEA), were used to quantify the energy consumption in buildings during the operation phase and its related environmental impacts. This life cycle quantification was addressed using diverse approaches, e.g., assessing the optimal insulation conditions, the management of heating and cooling systems or the influence of human behavior. These studies noted the importance of integrating the energy design of buildings for adapting to the user's lifestyle and optimizing the energy demand. Additionally, LCA and LCCs evaluated the benefits of building materials and maintenance activities in terms of improving energy efficiency.

The Net-ZEB concept is an extension of the NZEB, which implies that buildings cover their energy demand using onsite renewable energy sources (strategy B2). This generated a growing number of studies that integrate renewable energy into the total LCEA of buildings. LCEA and LCC are often used to analyze the effects of installing such systems. Results vary depending on the climatic conditions, the materials included in the energy equipment and the high cost of producing this equipment.

After optimizing the operation phase, using more sustainable construction materials is essential (B3). Many studies focused on the embodied energy of building materials, constructive systems and buildings using LCA. However, renewable insulation materials still need to undergo sufficient development to be comprehensively implemented in the building sector. Renewable materials such as kenaf fibers, cotton, jute, flax, hemp and cork were included in environmental studies, but more environmental and economic data on manufacturing is needed.

To increase the durability of buildings (B4) different alternatives were studied. Refurbishing buildings might improve their energy efficiency and increase their lifespan. Often, these improvements consist of increasing thermal insulation. Further, this strategy was partly applied by incorporating deconstruction materials into buildings, which increases the durability of building materials and facilitates future reuse.

So far, LCA has been widely applied in the building sector from an environmental perspective; in contrast, economic assessments are limited. Some specific studies addressed social aspects in buildings, but more efforts are needed to understand the social dimension and the effects of buildings on users, who play a key role in the implementation of energy efficiency strategies. Furthermore, most of the studies included a complete set of environmental indicators, while some others focused on the energy performance. Finally, there is also a need for internationally agreed data and methodologies for conducting these analyses at different scales.

Energy

The transition towards renewable energy (E1) was repeatedly reported in sustainability planning. Although renewable energy was widely assessed from a life cycle perspective, only a few studies focus on urban energy systems. The generation of domestic hot water was studied, concluding that geothermal energy generates lower environmental burdens than solar thermal. Another system are micro-wind turbines for electricity generation, whose environmental burdens depend on the geographical location and material composition. Researchers presented practical design principles for integrating renewable energy into the built environment. These studies lead to implementing self-supplied energy resources in cities (E2).

In the framework of urban parks, researchers assessed global self-sufficiency by considering factors such as the potential production of renewable energy. Both approaches contribute to increasing energy efficiency in cities (E3). At the urban level, researchers studied two types of street lighting using LCC (e.g., high-pressure mercury (HPM) and LED). Whereas HPM costs depend on the operation phase, the installation phase is more relevant in LED technology. The best option will therefore depend on the cost of energy in a specific urban area.

The assessment of energy issues relates to other urban issues, such as buildings (B), mobility (M) and urban planning (UP). In this section, only those studies associated with energy planning and production at the district and urban scales were included.

Accordingly, more studies are needed to evaluate the performance of renewable energy towards urban self-sufficiency from a sustainability standpoint, particularly considering the economic and social dimensions.

The latter would be beneficial to address the access to energy in different types of cities and social strata. Although researchers applied "LCC affordability", the income of citizens was not addressed to evaluate the access to energy (E4). Regarding specific methodological aspects, some studies limited the assessment to the global warming (GW), although such practice can lead to trade-offs among impact categories, as energy has a large influence on the GW.

Food

Urban agriculture was recently approached through life cycle tools and few articles are available. Although studies initiated in 2010, life cycle data was not experimental until recently. Some authors evaluated the potential environmental benefits of urban agriculture at the city scale. To do so, they accounted for the environmental savings resulting from local food production, which replaces the consumption of conventional food and, thus, avoids the environmental burdens associated with their supply-chain. Only a few types of urban agriculture forms were assessed in the literature, e.g., urban gardens, rooftop greenhouses and rooftop community gardens.

However, researchers employed literature data on conventional agriculture as a proxy for evaluating urban agriculture. Researchers combined LCA and LCC to observe the environmental and economic performance of rooftop agriculture. The social dimension of urban agriculture is yet to be considered in life cycle studies. In general, papers highlighted the benefits of urban agriculture as local production initiatives and demonstrated the potential environmental and economic savings from a consumer perspective.

Although research on local production does not exclusively refer to cities, urban areas are the main food consumers. Several LCA studies assessed local supply-chains by accounting for the environmental burdens of local and imported products. Most studies considered local production in Europe (e.g., UK), with apple being the main product of interest. Although local production is commonly associated with lower environmental impacts, results depend on the geographical area, cultivation technique and production stage. Morgan (2009) emphasized the need for using LCA in the evaluation of local food to include all the life cycle stages. The economic and social burdens of local production were not evaluated in life cycle studies.

Regarding the methodological approach to urban food, some limitations exist. Studies on urban agriculture mostly focused on the environmental dimension. In particular, attention was paid to the environmental savings of local production by calculating the avoided burdens of replacing conventional food with local food. Further environmental, economic and social benefits should be considered in comprehensive life cycle studies. Additionally, some of the existing articles are based on literature rather than experimental data from real urban crops. Furthermore, studies must represent the plurality of urban agriculture forms, cultivation techniques and vegetables. In terms of geographical representation, most of the papers assessed European case studies, unveiling the need to consider other regional areas for increasing the global knowledge. For both F1 and F2 strategies, studies focused on two indicators, i.e., GW and energy consumption, due to the importance of transportation in these environmental impacts. However, Payen et al. (2015) demonstrated that trade-offs can be produced among indicators and, thus, it is recommended to include a set of indicators rather than a single one. Finally, the economic and social dimensions of sustainability are commonly avoided in life cycle studies of urban food production.

Green spaces and landscape

Two analyses evaluated the life cycle performance of green areas in cities (G1). Researchers accounted for the carbon footprint of an existing green space project and determined that the benefits of carbon sequestration might be notable, especially when reducing the maintenance intensity and tree mortality. From another perspective, researchers highlighted the environmental impacts resulting from services provided in urban parks (e.g., museums, sports centers, or recreational areas). In the field of urban green areas, many studies focused on the carbon storage and impact offsetting related to trees, forests and natural green areas. However, integrating these results into the global environmental performance of green areas is needed to identify the balance between the life cycle impacts and benefits.

Regarding the implementation of green elements in buildings, studies assessed green roofs and green walls (G2). In particular, green roofs have multiple functions in cities and, thus, they can also be evaluated as insulation systems or water management strategies. In this sense, authors mostly considered collateral effects in their life cycle studies, particularly carbon sequestration and improved thermal insulation. From an economic standpoint, the monetization of these benefits highlighted the potential of green roofs compared to conventional façades and roofs (Porsche and Köhler, 2003). Regarding the design, using recycled materials might reduce the environmental footprint of green roofs and the maintenance of the system was a key contributor to the economic and environmental costs. However, most of the studies did not include a comprehensive life cycle approach that considers all of the stages. Researchers covered this methodological gap by conducting complete life cycle analyses of green roofs. Nonetheless, as green areas are multifunctional systems, studies should consider a complete set of impacts and benefits to estimate their real contribution to urban sustainability.

Regarding the social dimension, researchers unveiled the need to address the social impacts of green roofs and proposed a S-LCA framework, although it only discusses potential indicators related to stakeholder evaluation. In terms of indicators, most papers account for various impact categories. Evaluating a single environmental benefit of green roofs might underestimate their role in the design of sustainable cities.

Mobility

Some authors assessed the reduction of private car usage by promoting alternative ways of transportation (e.g., bicycle, public transport, walking) (M1). From a life cycle perspective, researchers reported positive environmental consequences resulting from the use of bicycles and/or public transportation when replacing private cars. A complete sustainability assessment was provided by researchers for comparing public and private options, where public transportation scored better in the environmental dimension, but needed improvement in social and economic terms. However, this study did not consider the life cycle approach.

The efficiency of private vehicles was covered in strategy M2, which consists of promoting the purchase of efficient cars or applying higher taxes to high-polluting vehicles. In this sense, different types of technologies were assessed using LCA, such as electric and hydrogen vehicles and propulsion technologies (e.g. conventional fuel engines, low sulfur diesel engines, compressed natural gas, or electric vehicles). Further, researchers studied the environmental performance of public charging facilities, which might be combined with solar panels for better results. Public transportation is also involved in the use of greener vehicles (M3). Different types of fuels were assessed in buses (e.g., diesel, natural gas, or hybrid options). In this case, the authors could not identify the best solution given tradeoffs among the impact indicators considered. However, as detected in other disciplines, an LCA of this type of strategy might identify the potential decentralization of pollution, meaning that cities might import or export environmental impacts through commuters.

The logistics sector is also involved in urban mobility. Regarding heavy trucks, the creation of a new distribution system that concentrates the delivery of commodities in the outskirts of towns is related to strategy M4. To the authors' knowledge, no studies dealt with this option so far, but they might be necessary for determining the balance between the investment in infrastructure and fleet and its environmental benefits. Strategy M5 referred to the use of ITS, which provide information to drivers about the best routes, for instance. ITS have a great potential to reduce CO_2 emissions resulting from private transportation. Nevertheless, this study focused on ITS benefits in roads and not in cities, and further research is needed. Additionally, future studies should determine the potential benefits and social acceptance of using ITS technologies in cities.

In the field of urban mobility, studies need to validate the sustainability of each strategy quantitatively. A life cycle approach might be useful to track the processes involved in the transportation sector. In addition, the elaboration of eco-efficiency (e.g., LCA+LCC) indicators may be interesting to determine the relationship between the money invested by public administration and the resulting environmental benefits. Existing studies already coupled some economic and environmental indicators, but the environmental dimension should include a wider set of indicators. Finally, S-LCA studies are lacking and they are needed to estimate the impacts of transportation on the society. Furthermore, future studies may include the influence of social participation on the overall acceptance and application of these strategies.

URBAN CHALLENGES

Cities operate in a resource-constrained platform that is full of challenges as they work with land use and planning priorities. Among the different ways to structure the challenges of cities, one conventional approach is to categorize issues by type of city department responsible for addressing them. Thermodynamically, challenges in our cities can be linked to the quantity and quality of material and energy generated and/or consumed in cities. These material and energy flows affect land use in and beyond the city boundaries.

Creating a sustainable city is ensuring how it deals with its material and energy flows and how its physical assets are developed and managed from a life cycle perspective. It is about realizing ways of developing and managing sustainable buildings, water supply, waste management, wastewater treatment, and transport systems. It is about how a city serves its residents and how its economy is managed. The challenges are presented here in terms of material and energy flows entering, occurring, and leaving cities.

Material aspects of cities

The material dimension of cities includes the buildings, the infrastructure, green spaces and parks, and the transport. Under infrastructure, roads, city rails, water infrastructure, and wastewater infrastructure are covered. It covers both bound materials and material flows through cities. From concrete to steel, huge amounts of materials reside in and pass through mega cities. The amount and type of materials in the building and infrastructures has been accumulating without meaningful thought concerning their fate. Traditionally, there were no regulatory requirements or economic or other forms of incentives to force or encourage material specifiers, builders, etc. to account for the end of life of the construction materials after the service life of the building or infrastructure is over. The toxicity buildup in these materials is worrisome, should the emitted substances make their way to our surface and groundwater or to the atmosphere.

When human beings used materials to build in the early days of urbanization in Asia and Africa, the materials were dominantly monotype and were loosely connected. Biomass in the form of straw, wooden beams and lintels, sun-dried clay mud, and stones consisted of most of these building materials. The materials, if necessary, could be disconnected easily and reused more than once. These materials were also harvested and collected from the immediate environment locally. With the advent of synthesis chemistry and the introduction of composite materials, the complexity of construction materials has increased significantly. The subsequent construction technologies of meshing up different types of materials have led to inseparable end products that can no longer be reused in a second life. The distances from which materials are sourced have also increased significantly for most of the supplies.

What we do in cities today will be critical in terms of influencing the material content of the existing building and infrastructure stock, which will stay for the next 50 years or more depending on the life time of the building and infrastructure. In the past and still in most cases, majority of these materials after demolition are sent to landfills. For lack of space and not-in-my-backyard pushes from local communities, cities are finding it difficult to get landfilling spaces to cope with the increasing quantity of building and demolition waste.

Regulatory restrictions and increasing awareness will force municipal decision-makers and developers to find new ways of sourcing new materials and using them in new constructions. Construction technologies need to account for the eventual disassembly of materials used in buildings and infrastructures and separation by material type.

Our knowledge of what materials contain and how they should be connected and used in buildings and infrastructure systems should account for how they will be disposed of at a later stage. Cities can save on extraction of virgin raw material and avoid other resources consumed during the extraction, processing, and transport of materials.

Water and wastewater aspects of cities

Many urban areas around the world do not provide basic services of clean water supply and adequate sanitation. In 2015, 2.1 billion people lacked access to safely managed drinking water. Waterborne diseases are responsible for a significant number of deaths and lost productivity and shortened life expectancy in many developing countries. Almost 1000 days die of waterborne diseases associated with lack of appropriate sanitation and clean water before celebrating their fifth birthday.

Cities have been from the very beginning on a wasteful trajectory when it comes to water supply and treatment. One aspect of this resource leakage is that our urban areas continue to utilize potable water treated through a series of technologies and consuming energy and chemicals for treatment only to be used to flush toilets. We are so locked-in to this obsolete system that our plumbing education and building codes are tied to this unduly common and old inefficient practice. The dependence on these paths and system has prevented decentralized innovative solutions from making important contributions both in incremental and transformative changes of how water and wastewater services are delivered in cities.

The solutions that are proposed to areas where there are poor and no such services at all are often large and centralized systems. Due to increased urbanization, many cities are overstretched in terms of water and wastewater treatment plant capacity. Building new systems and upgrading these large systems require big investments. The amount of energy consumed in the treatment of water and wastewater and the transport and distribution is one hindrance in operating and maintaining existing systems in cities where topography is challenging. Nutrient leakage associated with wastewater management is another aspect that needs attention.

Solid waste aspects of cities

The solid waste problem has economic implications in many developed cities. In developing countries, the issue takes a multifaceted form as it affects the social, environmental, and short-term and long-term economic advancements of urban areas. With increasing urbanization, the composition and quantity of waste is surpassing the already meager resource allocated to proper management of the waste in many cities of the developing world. According to the UN, the share of the world's population living in cities is expected to rise from 54 percent in 2014 to around 66 percent in 2050.

The rate of increase of volume of solid waste generation is higher than the rate of urbanization as established by a World Bank report from 2012. By 2025, the planet will have 4.3 billion urban residents generating about 1.42 kg/capita/day of municipal solid waste, that is, 2.2 billion tons per year. This increase in solid waste, which is the single largest budget item in many municipalities, will create unprecedented stress in cities in many developing countries, which are already operating under capacity in properly managing their solid waste. If current trends continue, a badly needed global "peak waste" will not happen this century.

Energy aspects of city buildings and transport

The heating, electricity, and fuel consumption of cities dominates the overall consumption of energy by society. Cities in the cold climate zones of the world specifically account for higher consumption of heat. The hottest regions of the world consume significantly high amounts of ventilation and air-conditioning energy. Decisions about the type and quantity of fuels used to deliver the relevant energy services often revolve around their energy intensity and impact intensity. Cities in different countries therefore have legislated to limit the amount of energy consumed by buildings. There has also been an increasing tendency of moving away from fossil fuel consumption in heating and electricity generation. Those in the cold climate regions will have to plan how their buildings and infrastructure will be heated during the long winter months of the year. The long-life time of buildings and the associated lock-in effect of the existing building stock require innovative approaches to redevelopment and urban renewable plans.

As the power grid decarbonizes, many cities in the world will continue to deal with their transport fuel. The amount of fossil fuel consumed for transportation in cities across the world is still the dominant contributor in both energy consumption and associated emissions of different types. This can be attributed to the high-carbon fuels and the increased mobility and expanding urbanization-led infrastructure systems. The expansion of these systems contributes to increase in indirect consumptions of energy through increased economic activities triggered by the infrastructures.

Measures aiming at energy efficiencies may not necessarily lead to the desired outcome of net reduction of energy and associated impact at the society level because of a potential rebound effect. Any extra money associated with energy efficiency when spent on previously nonexistent activity comes with higher energy consumption such as long-distance leisure travel; any gain from the energy efficiency is offset by the new energy-intensive activity and undermines the otherwise positive energy efficiency program.

Legislative and incentive instruments aiming at reducing energy use in cities with focus on buildings and transport systems better take a bigger picture and look beyond the different subsystems and factors including technological, architectural, urban form, lifestyle, behavioral, and cultural aspects.

LIFE CYCLE-BASED SOLUTIONS FOR SUSTAINABLE CITIES

Many of the problems that are established as critical in the context of urban areas are characterized as severe based on partial considerations of the full life cycle. They can be worse when accounted in their totality from a life cycle systems' perspective considering the downstream and upstream systems as well. Cities at the same time have a capacity of leverage in moving society at large in the right direction. Dematerialization of the urban infrastructure, exploring the area of integrated infrastructure and life cycle-based performance labeling are discussed as important elements of such opportunities. Capitalizing on such opportunities, however, demands a paradigm shift in the way cities do business.

Dematerialization of urban infrastructure

Innovative designs can be sought to reduce the quantity of materials extracted, processed, and utilized without compromising the quality of the infrastructure. This reduction in the amount of materials is not limited to what is finally bounded in the urban form. A life cycle perspective provides the opportunity to take stock of the materials that are wasted upstream in the mining and quarry sites as well as the waste that should have been diverted from landfills in and around cities. The life cycle lens goes beyond the embodied material during the preuse construction phase.

Services provided by and on urban infrastructure should also be dematerialized. Dematerialization is better seen as covering both a relative and absolute decoupling of resource consumption and associated environmental impact from economic growth. It is realized through concerted efforts of achieving significant increase in material and environmental efficiency. At the broadest level, urban areas around the world should be developed and operated with an additional type of decoupling in mind that decouples human well-being from economic growth and consumption through a set of measures that include reduction of excessive consumption levels. Decarbonization as a specific case of dematerialization is best applied in the form of decarbonizing the energy and transport systems, which are the top two contributors for greenhouse gases in cities around the world.

Assessment integration and integrated infrastructure

Given existing building and infrastructure stock are here to stay for long, innovative technical and financial mechanisms are required to reuse old infrastructure systems and as last resort recycle materials from them. Repurposing should be a prioritized norm in renewal and redevelopment works of urban centers across the world. The task of repurposing will be increasingly huge with game-changing technical transformation globally. One good example is the large-scale introduction of autonomous vehicles. The need for our massive parking surface and above surface structures will be reduced significantly, and large part of these structures will be abandoned. It is, thus, time to rethink new purposes for them and redesign and redevelop them for a second life and beyond. New projects can be designed and implemented using advanced knowledge using our experience and by finding synergies between different urban forms. Best practices around the world should be emulated with focused attention to local contexts and variables. Urban decision-makers will benefit from twofold integration in relation to infrastructure systems in cities: integration of assessment aspects and integration potentials of infrastructure systems.

Under the assessment integration, elaborating the concept of a comprehensive integrated assessment in relation to decision analysis of infrastructure development and management is important. Setting higher environmental, social, and economic standard on our cities requires the integration of decision-making related to urban infrastructure systems by looking at the triple bottom line as a more comprehensive platform for complementary consideration of the most important and relevant variables while avoiding double counting of the aspects informed by such variables.

Understanding the sustainability (environmental, social, and economic) performance of current best practices of developing and managing the totality of urban infrastructure systems and how future technical and nontechnical changes influence the sustainability is crucial. Conventional practices focus either on only one aspect of the sustainability or on only one infrastructure at a time and/or only part of the life cycle of the infrastructure. Decisions based on such partial information and circumscribed knowledge should be replaced by more complete knowledge as an input to the decision analysis stage of the development and management of future infrastructure systems. A comprehensive account of relevant environmental, social, and economic aspects of our urban infrastructure provides municipalities and higher level of governments the opportunity to look at the opportunities, the synergies, and trade-offs associated with alternative development and operational decisions.

Our understanding of the performance of the current best practices of developing and managing infrastructure systems from environmental, social, and economic perspective is weak. Even weaker is the level of knowledge we have around the changes in performance in the future with the introduction of technical and nontechnical shifts that can potentially happen in the next 10–30 years. Without an actionable knowledge regarding the current and future potential performance of our cities, we will fall short of getting it right in terms of what measures will be critical in containing the negative impacts of disruptions that affect infrastructure systems from a life cycle perspective. Integrated assessment provides the opportunity of identifying cobenefits and adverse side effects, which will otherwise lie outside conventional assessment lens.

On integrated infrastructure, exploring how different levels and scales of physical integration, data integration, resource integration, management integration, and other forms of integration affects the overall performance of infrastructure systems is critical. In some cities, there is already a demonstration of integration of infrastructure systems through connecting material and energy flows of individual systems. Connecting the waste management infrastructure with the transport system using fuels such as biogas produced from waste to power vehicles is one example. Or the use of waste-driven electricity to power transport systems could link the three infrastructure systems through material and energy flows. Advanced integration will be physical (e.g., surface) integration. The future which in many areas is already here holds promising technologies in terms of integrated infrastructure such as solar shingles and solar panel roads.

The photovoltaic layered roads or pedestrian ways, for example, can increase the values of infrastructure by adding new functions or layering functions, which is the basis of life cycle assessment's focus on functions of product systems. Envisioning new vehicle technologies that can charge while driving wireless/wire free from the roads and even from other autonomously driven vehicles is not wild. These kinds of developments will, of course, require all kinds of new fiscal, regulatory, and social transformation to work effectively.

Life cycle-based labeling and certification

As cities and nations increasingly set greenhouse gases-focused goals and broader sustainability targets, there is a need for measuring, monitoring, and communicating performances and progresses made. Increased sustainability awareness of citizens and the demand for better options offered by cities and parts of cities to prospective residents would encourage the development of sustainability rated or certified infrastructures, districts, neighborhoods, and cities in the future. Assessments that can potentially support such ratings and certifications come in different shapes and sizes depending on the purpose, scope, and object of assessment.

Life cycle assessment is about evaluating the environmental impacts of product systems such as buildings and infrastructure systems over their life cycle. Environmental impacts covered in a comprehensive life cycle assessment include climate change, ozone layer depletion, smog, eutrophication, acidification, human toxicity, ecotoxicity, biotic resource depletion, abiotic resource depletion, and fossil fuel depletion. Life cycle sustainability builds on life cycle assessment to include life cycle costing and social life cycle assessment. Metrics for monitoring and communicating supported by quantitative data are helpful. There is, however, a need for scrutinizing the quality of data, for example, by qualifying the assessment results and conclusions. More and better data collection from primary sources will be useful in addressing the kind of uncertainty associated with, for example, urban greenhouse gases inventory.

Verifiable and independently reviewed sustainability labeling and certification of buildings, neighborhoods, infrastructure systems, subcities, and cities has the potential of sending the right signals to the market and thereby fostering the expansion of best practices and state-of-the-art design, development, and management of the different levels of organization of our urban areas.

The performance labeling of infrastructure systems and other elements of urban forms will benefit from the experience of working with the life cycle-based environmental product declarations (EPDs) in the building sector. Rules that set the principles and requirements to be followed in developing EPDs are set in product category rules (PCRs) established through deliberative and participatory processes involving different stakeholders including relevant industrial associations. Both PCRs and EPDs as a basis for verification, comparison, and evidence-based monitoring are developed as living documents and are updated from time to time to capture new data and knowledge, new technology, and new requirements.

THE FUTURE OF SUSTAINABLE URBANISM

On the basis of the reviews of a history of sustainable urbanism and the literature on green cities, or eco-cities, six general principles for designing a sustainable city are defined:

- Close cycles at the lowest possible scale;
- Build redundancy in the design;

- Create anti-fragility;
- See citizens as (design) experts;
- Use the landscape as the basis for urban growth; and
- Develop innovative, rule-breaking designs.

Bringing the analyses of the history of sustainability in urbanism and the theory of eco-cities together, three comprehensive aspects of sustainable urbanism can be distinguished: the relation between society and the research community, the complexity of urban resilience, and the role of 'landscape' in urbanism.

Principles for Sustainable Urbanism

The six principles for sustainable urbanism are the following:

- Close cycles at the lowest possible scale. When the cycles of water, energy, and materials are closed, no waste streams are produced and the city is clean and has an efficient urban metabolism. The circularity provides abundant opportunities to create economic benefits. However, it is not always technically and financially most feasible to close cycles at the lowest possible scale. As researchers have pointed out, for energy solutions on different scales, up-scaling from the individual building can also lead to a balanced urban system;
- Create spaces for the unknown. When redundancy becomes a standard part in urban design, the city has the flexibility to 'host' unprecedented events. If, for instance, a cyclone hits the city, there is space to temporarily store the accompanying water. Researcher has identified this space as being 'unplanned';
- Design anti-fragile spaces. Using this new concept to design places that improve the quality in the city when an event occurs. 'Ant-fragility is defined as a convex response to a stressor or source of harm (for some range of variation), leading to a positive sensitivity to increase in volatility (or variability, stress, dispersion of outcomes, or uncertainty, what is grouped under the designation "disorder cluster")'. Applied in urban environments under stress of climate impacts, anti-fragility is the concept that urban environments become stronger than before through their response to climate events. In urbanism, this concept is ill-researched. However, it could potentially open new design pathways to create urban environments that not only could withstand or bounce back from climate disasters, but, as a result of the hazard, grow stronger;
- Let people own their environment. When residents are part of the design process, they will create ownership over the changes and invest in their

environment. Once residents own their environment, they will make sure it is maintained well. The design charrette is a wonderful design method to develop designs in an inclusive way.

- Use the landscape as the basis for urbanism. Landscape systems, such as the water system, ecology, and soil, create the conditions for urban occupancy. The values in the landscape then will determine the type of urban development. This is contrary to the current situation where the urban development dictates the landscape, e.g., overthrows the existing landscape values to create ordinary residential areas. Even in existing cities, the underlying landscape can be illuminated and brought to the surface to give the city extra ecological qualities.
- Create innovative designs. The current conventions need to be broken if a sustainable city is to be developed. The existing procedures need to be bypassed using creative design approaches. For the development of regional plans, metropolitan spatial visions, district plans, and neighbourhood designs, a sabbatical detour can provide the space to think differently and the time to come up with alternative design propositions. Elements of a sabbatical detour are design competitions, an idea-generating charrette, designers in resident, exhibitions, and many other activities.

Society Based, Complexity-Led, and Landscape-Driven

The six design principles are interrelated and can be coherently summarized under three comprehensive concepts.

Society-Based Urbanism

First of all, a key feature of an eco-city is a new relationship between Research/ Governance and Society. Recent developments in the history of sustainable urbanism and the literature on eco-urbanism show that society increasingly plays the role of co-designer. The knowledgeable citizen has appeared and he/she is no longer satisfied with a role as respondent to plans made by professionals and/or the government. This implies a different role for the designer himself. Instead of being the 'planner who knows', the designer must become the 'planner who listens'. Before a design for a city is made, the designer as well as governing bodies will benefit from the knowledge and local expertise citizens have and can bring in the design process.

The design then is enriched with additional understanding and emotional connotations of the city at stake. The research agenda therefore is not only determined by the funding agency, or the brains of the researcher, it will also be influenced through input from society.

A recent planning process undertaken within the frame of 100 resilient cities in Sydney has revealed that citizens came up with approximately 400 ideas that make the city more resilient. Cross-pollination between research agendas and the civil agenda is necessary to grow a sustainable city.

Complexity-Led Urbanism

Secondly, the reviews have revealed that thinking about the city as a complex system is becoming more prominent. The different subsystems in the city, such as the transportation system, the energy system, and others, are increasingly seen as complex systems as such, but these feed in to the complexity of the city as a whole. The complexity of the entire city is difficult to grasp, as the interrelations, dependencies, and connectivity between all subsystems in the city are complex by nature. Earlier research therefore proposed to deconstruct urban complexity in 15 systemic layers.

Amongst these urban systems are the zero-carbon city, the climate resilient city, the densified city, the inhabitable city, the affordable city, the healthy city, the metabolic city, and more. Once these 'single-issue cities' have been designed to their ultimate, most resilient shape, they can be recombined. Then, the complexity can be reconstructed at the neighborhood scale, where local potentials and identities can be generated.

Combining the specific features of the different layers for each location gives these places a very specific identity distinct from their neighbors. And, because the reconstructed complexity is based on the resiliency of individual layers, the overall capacity to adapt will increase both at the local level and for the entire metropolis. Because the set of 15 layers contains technical, socio-economic, and nature-based aspects, the combination of these layers provides a balanced and integrated way of looking at the entire city.

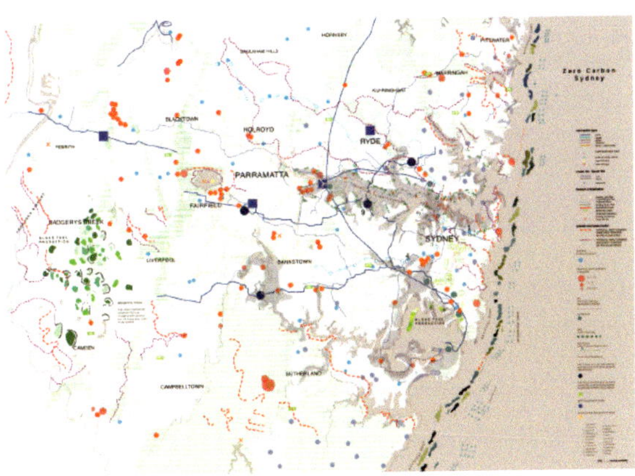

Fig. 1. Map showing the Metropolis of Sydney as a Zero-Carbon area

Research on these 15 layers is elaborate and the set-up of a rigorous data-base is time-consuming. Recently, investigations into the Zero-Carbon Metropolis of Sydney has led to an image of the first layer (Fig.1), which already can form the basis for further resilient designs at a lower scale. This first layer has been developed by mapping the metropolitan renewable energy resources of the Greater Sydney area. These mappings of the wind potentials, hydropower potentials, solar potentials, and biomass potentials have been connected with the demand of households, public utilities, and industries under an ecological 'zero-carbon' lifestyle scenario. The integrated map shows the combination of these types of renewable energy supplies.

Landscape-Driven Design

Landscape is everywhere around us, but in recent decades it has not played a dominant role in the growth of cities. The landscape is often neglected and replaced by an urban landscape that marginally contains elements of the former landscape. Both reviews illustrate that the landscape, represented through its elevation, water, ecology, is a crucial factor to create more sustainable cities. These factors, if not used as a determining factor in urban development, will eventually counterpunch at the city in the form of severe flooding, absence of agricultural fertility, or other disruptive processes. When these factors, however, are used as the basis for urban development, a truly sustainable city could emerge.

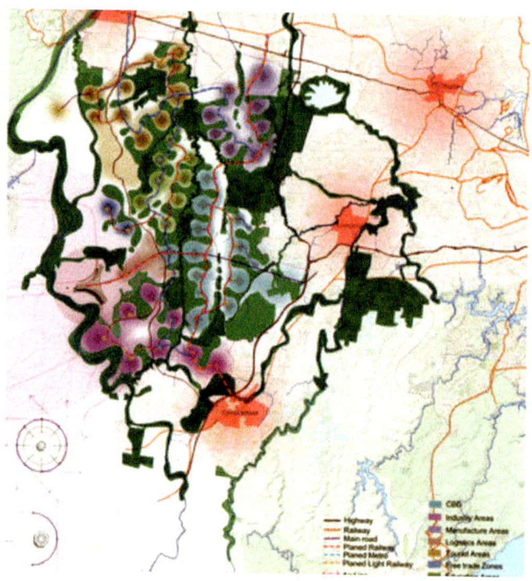

Fig. 2. Plan for a landscape-driven urban design for Badgerys Creek

The processes that drive the landscape are then also the drivers of urban development. If the flows of water form the basis for the urban lay-out, or if the connected habitats of natural ecologies are taken as the starting point of urban development, or if the natural gradients of higher oligotrophic grounds and the lower eutrophic soils are kept intact, the city that emerges builds on the natural richness of the landscape, creating a city in harmony with its environments. Recently, studies have been undertaken for the Badgerys Creek area, west of Sydney, where a new airport and a city with one million new inhabitants are planned to arise. If the urban growth is guided by the principles of the landscape, a very interesting and attractive urban environment can be designed (Fig. 2).

REVIEW OF URBAN DENSITY AND COMPACTNESS

Urban density and mixed-use are key factors in determining the sustainability of a precinct or neighbourhood and its urban liveability. Mixed-use neighbourhoods are more likely to offer employment locally. Urban districts have a significant complexity about them, and clearly there is still a need for more research, comparative data and an evidence base on the benefits and detriments of more dense and compact cities, which has frequently been noted by different scholars.

Urban density is a term used in urban planning and design to refer to the number of people inhabiting a given urbanized area, and the amount of floor area built on a defined site. It is considered an important factor in understanding how cities function. However, the link between urban density and aspects of sustainability remains a contested and often misunderstood subject of planning theory. And residents do not know enough about densities in cities, but are concerned about potential negative impacts.

Remembering hyper-density: the case of Kowloon Walled City in Hong Kong

Poorly managed (or neglected) and badly planned density can lead to overcrowding, overdevelopment and lack of daylight, as was the case with the Walled City. For a long time Hong Kong's extremely high density has led to apartments which have never seen a ray of sunshine and to narrow streets described as 'airless canyons'. The lack of natural ventilation and cooling breezes in the city at pedestrian level has increased the urban heat island effect and brought serious health impacts by changing the urban microclimate.

Ungoverned, uncontrolled and unregulated, the Walled City was a huge block of around 300 interconnected buildings ranging from 10 to 14 floors, on a small site of only 2.2 hectares. According to Ian Lambot's documentation in 'City of Darkness', Kowloon Walled City began as a fortress used by the Chinese to defend the city against the British invasion in the middle of the nineteenth century and later served as a hiding place for gangs and criminals. An amazing labyrinth of passageways, small rooms and courtyards, it contained a great variety of functions: a mixed-use, hyper-dense block that was home to around 25,000 families and businesses living in high-rise buildings, frequently in windowless cramped flats, and all constructed without a contribution from a single architect (Figs. 3a, b, c, d).

Opium parlours, prostitution and unlicensed dentists alongside kindergartens, tiny factories and food stalls – the infamous Kowloon Walled City was once one of the densest places on earth. With a population density of 1.2 million per square kilometre, the settlement had no government enforcement from the Chinese or British. The population burgeoned with refugees from mainland China and the area became a haven for vice, controlled by triads. 25 years have passed since it has been erased from the Hong Kong landscape and converted into a park, while some hold onto the memories of such extreme conditions of population density. As well as being the location for Hollywood films, such as Crime Story, the infamous settlement has even been recreated as a theme park.

Fig. 3 a and b: The Kowloon Walled City in Hong Kong, demolished in 1992-93 due to its extreme density and unhealthy living conditions. Left: aerial photo. Right: All that remains today of the Hong Kong legend: the bronze model of Kowloon Walled City in the middle of a public park where it once stood. c: Aerial photo of Kowloon Walled City, a compact superblock with over 50,000 residents. d: Kowloon Walled City in Hong Kong; it earned its Cantonese nickname 'City of Darkness'

Li Shiqiao wrote that Kowloon Walled City was associated with anarchy and lawlessness; it was 'the extraordinary example of a physical expression of the city of maximum quantities in an embryonic form'; and he notes: 'It failed because its quantities were never managed by expert knowledge of orderly planning, hygiene and safety'.

Good urban design always ensures sufficient daylight for residents and efficient natural ventilation of the spaces between the buildings and rooms inside. This extreme example of hyper-density, an over-densified superblock, could only develop in Hong Kong or in India with a culture of greed and powerful developers, where architecture is reduced to commodity. But it was also a result of the building code: in the 1960s and 70s, the inner-city airport Kai Tak caused a height limit to be set for buildings at 60 metres above principal datum for any development in Kowloon, which led to the typical Kowloon block: a very compact hyper-dense block of 12- to 14-storey buildings. Very dense and intense high-rise cities tend to be overcrowded and are not the best option. While the Walled City was a functioning tight-knit urban community (and a hotbed for crime), it was not sustainable and the negative impacts of too much density slowly made its residents sick.

The compact city revisited: recent 'superblock' housing experiments in Singapore, Sydney and Vancouver

The following section describes the selected cases: new housing experiments in compact superblocks in Singapore, Sydney and Vancouver that have introduced higher density, each in their own way. Sustainable urban design can include 'high-tech' design solutions as well as 'low-tech' alternatives.

The functional and organizational aspects of sustainable urban design focus on building structures, context-specific issues (such as site and local climate, regional architectural traditions and typologies, context, local building materials, etc.), solar energy, and the recycling and re-use of the existing building fabric.

Some of today's thinking around high-density precincts in China, Singapore or Sydney could translate to other cities and inform new approaches to increasing inner-city density and reducing urban sprawl, while promoting a sense of community. In the US and in Australia, change is happening too: while many residents in suburbs simply live too far away from their jobs, in the majority of North American and Australian cities inner-city populations are predicted to rise, especially on redeveloped brownfield and urban renewal sites, as people want to live closer to the city centre and in the inner suburbs, to reduce their need to commute or look after a large garden.

When you are densifying an area that is already dense, there is a question as to whether the existing infrastructure can cope. One other challenge cities face in their densification strategies is a reluctant public and resistance from residents against higher densities, as illustrated by the case of Vancouver's protesting neighbourhood groups. If done well, higher density does not necessarily decrease liveability (as can be seen in cities such as Singapore, Barcelona and London). Higher density living can be acceptable for residents as long as these developments also provide for new treed parks and urban greenery; but a well-used park a drive away cannot be a substitute for new green space. High-quality urban design can alleviate negative perceptions of density at the metropolitan scale. Higher densities require new better housing typologies, a wider range of compact housing models and innovative design solutions that integrate urban greenery and high-quality public space. Landscaping, green roofs and the design of community spaces must be important elements from the outset of each development.

The following selected cases deserve a closer look, because they have successfully introduced denser housing models and tested innovative typologies for urban precincts, where buildings and urban greenery, and public and private spaces, have been combined and intertwined:

- The Interlace and Pinnacle housing developments in Singapore's west,
- The Central Park development in Sydney,
- The False Creek development in Vancouver

These cases are compact and spatially complex models, featuring medium- to high-density housing typologies, with a fine grain of diversity and complexity, creating a 'vertical city' and matrix of horizontally stacked urbanism that incorporates communal courtyards and double-height balcony spaces.

The Pinnacle and the Interlace in Singapore

The city state of Singapore has a tropical, humid climate all year round, which makes ventilation and cross-ventilation the most important aspects of environmental comfort. The urban design principles that apply to cities in tropical conditions are different from these in temperate conditions. An effective way is to combine active and passive systems to radically improve the environmental performance of buildings.

Passive systems include building orientation, compact geometry (a reduced façade area reduces the solar heat gain), strategic positioning and sizing of windows for maximum cross-ventilation, thermally activated concrete slabs for cooling, effective shading, high wall insulation, green roofs and the integration of urban greenery.

After the success of the Pinnacle at Duxton (the world's highest public housing complex), the government-directed Housing Development Board (HDB) decided to go one step further and experiment with an even more radical housing development. The Interlace is one of the largest and most ambitious residential developments in Singapore, and it presents an approach to contemporary living in a tropical environment by adopting a new residential typology which breaks away from the standard vertical HDB tower blocks of Singapore. The Pinnacle is the world's highest public housing complex, consisting of 7 blocks, each 50 storeys high and connected by sky gardens and bridges on the 25th and 50th floor. The roof garden is an expansive area with recreational facilities for residents.

Up to now, most housing in Singapore has been in unsightly generic tower blocks, but the Interlace is an intertwined 'vertical village' where the apartment blocks are stacked in a hexagonal arrangement around eight courtyards, creating the impression of open space and fluidity. One could say that the Interlace introduces a new typology that could mark the end of generic tower blocks in Singapore. The large-scale complex also provides a more interconnected approach to living through communal spaces which are integrated into its lush surrounding garden (by introducing extensive roof gardens, landscaped sky terraces, cascading balconies and lush green areas).

Thirty-one apartment blocks, each six storeys tall and identical in length, are stacked on the eight-hectare site in a hexagonal arrangement to form eight large open and permeable courtyards: the stacked formation allows light and air to flow through the ensemble and surrounding landscape. While from the point of view of shared communal space it is much better than a tower, the disadvantage is that quite a few of the apartments are always under the shadow of the block directly on top.

The Interlace covers 170,000 sqm of gross floor area and houses 1040 apartment units of varying size, accommodating over 2500 people. The continuous landscape is also projected vertically, from the planting of green areas in open-air basement voids, through balconies and rooftop gardens. The private balconies give apartments large outdoor spaces and personal planting areas that look like cascading gardens, and sky gardens provide panoramic views.

The design also incorporates sustainability features through careful environmental analysis of the sun path, wind direction and microclimate conditions on site and the integration of low-impact passive energy strategies. Water bodies have been strategically placed within wind corridors as a means of allowing evaporative cooling to happen along the wind paths, reducing local air temperatures and improving thermal comfort in outdoor recreation spaces. (Figs. 4a, b; 5a, b)

The Central Park development in Sydney, Australia

Sydney's Central Park is a major mixed-use urban renewal project located on Broadway in Sydney's centre, very close to Central Station. Building heights range from 8 storeys to 35 storeys. The development is focused on a new public park located just off Broadway of approximately 6500 square metres (70,000 sq ft) in size.

The first stage of the redevelopment (completed in 2013) is a tower called One Central Park, a 117-metre-tall residential tower featuring 'vertical gardens' and a shopping centre in its podium on the lower levels. The design features a cantilevered section including a heliostat to provide reflected sunlight to the parkland and roof garden below. The precinct has a population density of around 1000 people per hectare. (Fig. 6a, b).

Fig. 4 a and b: The housing development Interlace, in western Singapore. This superblock project houses around 2500 people

The False Creek development in Vancouver, Canada

The Canadian city of Vancouver has a long-established reputation as an urban laboratory for successful experiments in density, downtown living and a strong neighbourhood community spirit. Vancouver's urban transformation and density debate over the last two decades has led to some much celebrated medium-density neighbourhoods in waterfront locations. Southeast False Creek was developed on a brownfield site and has often been criticized for its unaffordable housing prices and 'green overcrowding' (density without amenity). Nevertheless, this development and the wider urban renewal of Vancouver have been a success story: in the past 15 years, 60,000 people have moved to the city's downtown peninsula in medium- to high-density developments; and today, over 40 per cent of dwellings in Vancouver are apartments (this is approximately double the percentage in Sydney).

a

b

Fig. 5 a and b: The earlier housing development the Pinnacle, at Duxton in Singapore's west, forming a massive 'wall'.

Since its founding in the 1970s, False Creek South's residents have quietly enjoyed its view of the changing skyline of downtown Vancouver, with incremental increases in density spread over a decade. False Creek South was once a conscious experiment in neighbourhood-scale urban design, since studied and applauded by planners and architects from around the world. Just like in Sydney, as the entire city wrestles with high-value waterfront real estate, skyrocketing housing costs and development pressures, False Creek has been getting densified, and it has been able to absorb some additional density and accommodate some growth without losing the character of its neighbourhood. For instance, False Creek became the location of the athlete housing as part of the 2010 Olympic Village.

Fig. 6 a and b: The new Central Park development in Sydney, a high-density inner-city mixed-use development close to the Central railway station and in walking distance to three universities (completed in 2014). This is one of Sydney's new high-rise precincts with a population density of 1 000 people per hectare.

The City of Vancouver has plans to see this liveable neighbourhood further developed into a medium- to high-density residential area with housing and services for 11,000–13,000 people. The early parts of the development that date back to the 1970s resulted in a medium-density area with a variety of architectural designs, ownership opportunities, recreational activities, and access to modes of transportation. The original precinct has a modest population density of around 120 people per hectare. Bike paths, parks, unique three-storey homes, a public market and the intentional preservation of mountain views distinguish this area of False Creek as one of the earliest conscious attempts to create a liveable medium-density environment, rather than focusing on high-rise efficiency and profitability. However, more recently, the scale has shifted to 6- to 10-storey apartment buildings, and even to 20-storey towers, all located in walking distance to the downtown area.

With a population density of over 5 400 people per square kilometre, Vancouver is the most densely populated Canadian municipality, and the fourth most densely populated city in North America, behind New York City, San Francisco and Mexico City. Urban planner Wendy Sarkissian [64] is more critical of Vancouver's 'EcoDensity' policy and argues that strong community concerns and established policies of community engagement were simply ignored by the City Council in the move towards implementation of its top-down EcoDensity Charter. The EcoDensity policy was widely unpopular, and Sarkissian adds that the Laneway Housing initiative, which predated the EcoDensity policy (and which was much more about gentle small-scale densification), failed to deliver a significant increase in density. She argues that the real reasons for EcoDensity were prodeveloper and ideological. However, it did not provide affordable housing as promised, and instead drove housing prices up further, enabled more large-scale types of developments and served developers' interests (Figs. 7a, b).

Fig. 7 a and b: The development around False Creek in Vancouver, British Columbia, hasresulted in a compact neighbourhood with a mix of densities, from mid-rise 3-storey buildings; however, recently, the scale has shifted to 10- to 20-storey apartment buildings

CHAPTER 5

AESTHETIC ESSENTIALS

The aesthetic has been one of the important subjects for most philosophers, artists, and architects for centuries. Cities carried in the stages of growth various aesthetic values according to their locations, reasons of growth and the different goals of expansion and development, with the acceleration of cities excessive growth and urban sprawl, some cities begun to lose the basic or acquired aesthetic values, while others directed contemporary and future urban development toward supporting and maintaining them. This research examines the aesthetic values of the future cities and defines achieving and supporting them in designing and planning levels.

THE AESTHETIC VALUE

According to Oxford Dictionary, the meaning of the word aesthetic is "knowledge derived from the senses". In the sense 'relating to perception by the senses', and Webster Dictionary describe it as a branch of philosophy dealing with the nature of beauty, art, and taste and with the creation and appreciation of beauty.

According to Lang/1987 the science of aesthetic is concerned with identifying and understanding the factors that contribute to the perception of an object or a process as a beautiful or, at least, a pleasurable experience, and understanding the nature of human ability to create and to enjoy creating displays that are aesthetically pleasing.

The need for aesthetic is one of the basic human needs (utilitarian, symbolic, aesthetic), where its job is to satisfy the individual psychological requirements to enjoy the presence, accordingly the beauty can be defined as the individual's sensual value to the things, which achieve delight by the interaction of psychological sensory abilities with, by this pleasure and enjoy, it gives the psychology of self-value its existence.

The aesthetic judgment is concerned with all feelings, senses, and emotions in one moment, and has a relationship with physical actions, so aesthetic judgments are subject to cultural condition in some areas, the judgments on aesthetic value can be linked to economic, political and moral values, and so in terms of relying on feelings, emotions , mental, intellectual, views, the will and desire, in addition to culture, preference, behavior of non-conscious, conscious decision, training, instinct and social logic, or some of the other complex stuff and relying on the theory that is adopted.

Aesthetic considerations form undeniably one of the important aspects of urban experience - aesthetically satisfying environments, for example, increase human well-being. But what could the role of aesthetics be in attempts to create a more sustainable urban life? Some aesthetic decisions in an urban context can, for example, lead to an excessive use of natural resources and, thus, be in conflict with the goal of sustainability.

However, it is possible to look at the relationship of aesthetic considerations and sustainability from another angle. Innovative environmental art can be a powerful means to improve people's ecological awareness, thus forming one platform where ecology and aesthetics meet. The preservation of cultural heritage in an urban setting central to the issue of cultural sustainability has an aesthetic aspect to it as well.

The issue of the cultural sustainability of landscapes has also received growing interest. Moreover, the important role of aesthetics in urban well-being is hard to deny. The ultimate contribution that environmental aesthetics can have on the issue of urban sustainability, however, demands much more reflection.

Interaction between man-made environment and nature is necessary to attain sustainable development. In an urban setting, aesthetics is an essential requirement for humans, almost comparable to survival needs. Aesthetics in urban design may aid in building connected and coordinated urban growth that is built around mass public transport which creates economically dynamic and healthier cities. This also helps in the reduction of carbon emissions. On the other hand, an aesthetically unpleasing environment leads to a widening gap between people and nature.

It pushes people to avoid walking and enjoying the environment and encourages and increases cars (a leading contributor to the growing carbon emission in the atmosphere), providing them with door-to-door services. Therefore, an aesthetically unpleasing environment isolates the people and segregates them into economic caves and land use enclaves. It doesn't allow interaction.

Thus, to work towards sustainable cities, it is imperative to incorporate aesthetically pleasing elements into the environment. This section will first highlight ways of achieving this through design principles and will further focus of practical ways of achieving this in the environment.

DESIGN PRINCIPLES FOR ACHIEVING SUSTAINABLE CITIES

An aesthetically pleasing environment is necessary to provide an experience of nature to the people. It also creates a sense of wellbeing, positivity, social and psychological benefits and environmental and ecological services, which help enrich the built environment (Chiesura, 2003). A good physical and functional structure includes the design principle of variety, permeability, legibility, robustness and visual appropriateness.

VARIETY

Mixed-use development is a type of urban development that blends residential, commercial, institutional, industrial and cultural uses. Mixed-use development along the city's main axis, which has pedestrian and vehicular connectivity, will provide people with a better option for employment and housing. In addition to that people will enjoy the benefits of good connectivity using public and private transport. It will provide the residents with a better option of employment and housing in the area where the primary function is housing. It will promote walkability, thus supporting the 'eye in the street' concept. The presence of mixed-use development along the road will transform it into a safe street due to public activity throughout.

Fig. 1. Mixed use development.

PERMEABILITY

A way to achieve the aesthetics in the environment is through the design principle of permeability. Permeability makes spaces more inviting and approachable for the people. The walking experiences and street life becomes more enjoyable due to active edges. This design principle provides a good visual connection with the immediate surroundings and active impermeable boundaries. The surrounding of the pedestrian pathway should be porous to ensure safety and enhanced contact. One of the indicators for achieving this is to have doors to entrances directly visible from the pathways. This is beneficial for both the pass bys and people inside the buildings, achieved by having active and rich boundaries rather than passive and blank.

A good street connectivity shows more 4-way junctions and fewer dead ends or cul de sacs. Higher the intersections, the better the street connectivity. The lower number of street intersections proves a low street map connectivity and streets that are not functioning up to their capabilities. Therefore, it is imperative to extend paths and streets to nearby streets. This will overall increase the number of intersections.

Jane Jacobs built a theory of urbanism on the role of streets in her book '*The Death and Life of Great American Cities* (1961)'. The book stated that the most significant streets do not exist in isolation but derive part of their vitality from the surrounding street network. Areas with high connectivity allow easy movement in many directions and contribute to accessibility and ease of reaching places. In addition, a well-connected street network benefits the public, including direct routes and, therefore, shorter distances to destinations. It is imperative to create pedestrian links to shorten walking routes for making it for pedestrians. Good visual connection makes urban spaces approachable and keeps 'eyes on the street' from inside and outside the building. This ensures the safety of the people walking in the street.

Fig. 2. Good street connectivity.

LEGIBILITY

Robustness is the use of a place that does not have a fixed purpose and can be used for different purposes. The clarity of appearance of a place makes people aware of the choices available. Visual appropriateness also means the quick interpretation and identification of a space or area in an urban setting. An example of this is the easy identification of commercial and residential areas. On the other hand, richness is the choice of sensory experience and variety offered to users. An example of this is unique, exciting and contrasting building and urban design. Also, landmarks will be colourful, with different types of food, markets, venues, arts and gatherings.

Fig. 3. Legibility

ROBUSTNESS, VISUAL APPROPRIATENESS AND RICHNESS

Robustness is the use of a place that does not have a fixed purpose and can be used for different purposes. The clarity of appearance of a place makes people aware of the choices available. Visual appropriateness also means the quick interpretation and identification of a space or area in an urban setting. An example of this is the easy identification of commercial and residential areas. On the other hand, richness is the choice of sensory experience and variety offered to users. An example of this is unique, exciting and contrasting building and urban design. Also, landmarks will be colourful, with different types of food, markets, venues, arts and gatherings.

AESTHETICS

From the beginning of time, humans created different aesthetics through different periods depending on their interaction with nature. A branch of Philosophy, aesthetics is a nature of art that provides a meaning to life. Aesthetics may display in nature naturally as natural beauty or artificially as geometrical beauty through objects or forms that may have proportion, harmony or unity. Additionally, it may create a balance, identity, sense of place, and coordination. It goes beyond visual and physical characteristics and depends on human perception and multi-layer environment experience. Therefore, ecology and aesthetics are inseparable, making it a combination of qualities that charm intellectual or moral faculties. Aesthetics may also show motion recognising the flow of passing the time, demanding revolution, continuity and sensory experience. This generates cycles and awareness necessary to sustain life.

What makes a city aesthetically sustainable?

Lehtinen says that the aesthetic values pertaining to cities are also affected by cognitive, ethical and ecological values. What we find pleasant or aesthetically interesting always reflects, to a certain degree, our knowledge of the subject or phenomenon in question, and whether we otherwise consider it desirable. Not many would be so quick to admire even the most gorgeous sunsets in an urban setting if they knew that particle emissions generated by traffic have an impact on the intensity of the colours in the sky. Another thing that is important to understand is that aesthetic values constitute a broad spectrum of characteristics not limited to beauty. In a city, not everything can be purely aesthetically pleasing.

This means that aesthetic values are calibrated according to what is known about the functionality of various solutions. A good example is the renovation of the Alvar Aalto-designed Finlandia Hall in Helsinki, now in its early stages. In conjunction with the comprehensive renovation project, the Carrara marble used as the building's facade material, not well suited to Finnish weather conditions, will finally be replaced with a more durable option.

Sustainability and aesthetics are organically linked to urban architecture and planning: more consideration should be given to what will be demolished and how new buildings are built. Lehtinen believes that increasingly multidisciplinary collaboration is needed to investigate how technical challenges, the aesthetic layers of the city and the principles of circular economy are taken into account, in addition to costs, when making decisions about pulling something down.

According to Lehtinen, cities are the result of compromise, also aesthetically, since they manifest the preferences of several generations and parallel communities. Besides supporting contemporary urban culture, temporal and stylistic aesthetic diversity is what makes cities feel alive and pleasant to live in.

Aesthetics can contribute to urban planning

Philosophical and applied aesthetics has a lot to give to urban planning. Lehtinen thinks that we are living in interesting times, as Helsinki, for example, is busy with new construction sites from which entire new city districts are springing up. Construction is guided by value judgements.

The environment affects people even though people do not always have the chance to influence their environment. What matters, among other things, is what we see and how we move from one place to another.

"In terms of experiencing, it makes a difference whether you observe the city in your everyday life for a significant portion of time on public transport or behind the wheel of a car."

In addition to visual features, the city is associated with smells, flavours and soundscapes characteristic of it. This way, various food cultures and urban technologies become part of a multisensory urban experience.

THE AESTHETICS OF THE CITY

Jackson/1959 refered to three major goals for the design policies of cities (social-economic efficiency, biological and health), and pointed out that cities have to provide their citizens constantly with satisfactory sensory and aesthetic experience. That's why cities was considered according to the proposals of Mumford (The greatest artwork of human beings), and The built environment can represent a work of art at any dimension of human experience.

According to researcher giving the urban setting its full meaning, that requires aesthetic engagement which involve a visions learning experience from the natural, physical and emotional dimensions. As the aesthetic experience is not related only to build environment but includes also the living environments. Indeed, today's decision makers and developers do take aesthetic experience into consideration~ but it nonetheless contributes essentially to an aestheticization of the environment that favors a spectacular vision of the city.

Architecture is inseparable from life and society. People create them with an intention of aesthetic and the creation of better surrounding for life. This intention also goes with the creation of cities. Rossi thinks that 'urban history' is most useful to study urban structure. The continuity and therefore the history are important aspects underlying his theories. Urban aesthetics constitute a science founded in meaning inherent in the pre-existing building stock of the city.

AESTHETIC ELEMENTS OF THE CITY

The idea of aesthetic engagement involves an active experience as the aesthetic experience of environment increases the value of both the environment and the citizens at the same time. According to researcher the aesthetics of the cities lie in civil values, sharing of knowledge, cultural diversity and provide many means of transportation...in addition to provide means of cleaning the streets and removing waste, the geographical landscape, public spaces and facilities, rest areas and diverse music through local radio or itinerant music and strengthening the law which prevents noise and pollution.

The main vocabulary that increase the aesthetic value of the urban context is the diversity of natural vocabulary, as they are related to the aesthetic value of natural components of the physical characteristics such as color, shape, density and structure of horizontal and vertical.

Researcher noticed the importance of urban organization and coordination between spatial and temporal organization on one hand, and between the elements of urban configuration that meet the citizens and residents of cities' needs, these elements can be horizontal like roads and infrastructures or vertical elements of buildings lighting and trees..and so, as contemporary trends in planning and organizing call for the great attention to these aesthetic elements because of their inherent advantages and aesthetic quality and service life in cities and residential communities.

Researcher identified the basic aesthetic elements of the city as:

- Roads and alleys.
- Landscaping as an essential element in the physical configuration
- Elements of public lighting
- Urban furniture, functional furniture, aesthetic furniture
- Water in the city

Architecture and Urbanism.. Where the most important role in the Imagining of built sizes and the way they relate to and presence in a space, and then the implementation, as one cannot talk about art found in any form unless the architectural treated in his studies the following points:

- The proposed elevations and their compatibility with the existing buildings to avoid repetition, boredom and monotony.
- Road junctions and the buildings vista that forming them.
- The proposed perspective for the urban form and its impacts on the public spaces surrounding it.
- The proposed public spaces by designers as aesthetic elements.

NATURAL PROCESS INSTEAD OF NATURAL FORM

Natural form regeneration has been usually seen through the conventional method of ecological mimicry in the sustainable landscape. However, these are not sustainable as they require a high amount of maintenance and quickly become abandoned landscapes due to loss of stability. Therefore, mimicry of the natural process is more effective than mimicry of natural ecology.

Fig. 4 Duisburg-Nord Park.

An example of a regenerated natural process is visible through the Duisburg-Nord Park. The industrial pollution caused by waste coal and steel factories led to the site's abandonment for two decades. However, the ecological functions led to the consolidation of the new aesthetic attitude through a new canal with fresh rainwater instead of the old sewage canal. In addition, the canal spreads awareness for the visitors regarding water processes and changes in time

Beyond Ecological Performance

This principle highlights the beyond the performance of design regarding socially and culturally through events or festivals. An example of this is visible through Yanweizhou Park in Jinhua, China. The project led to reviving the disrupted river's ecosystem. Earlier the river was blocked with a protective wall to prevent seasonal river floods, disrupting the interaction between the inhabitants and the river. Removing the wall through a safe model and defining social and cultural objectives created social infrastructure and rebuilt society. The place is also used in the Chinese New Year festival because of the bridge connecting the two sides of the river, inspired by the local legends of dragons (Hemmti, 2016).

Fig. 5. Yanweizhou Park in Jinhua, China

Dynamic Instead of Static

Aesthetics in an urban setting is changeable and temporal. Aesthetics are variable and dynamic. For instance, not only do humans move in an urban environment, but the landscape moves, changing and declining.

An example of this is visible in Bass Riverbank, which was based on the design concept of dynamics. Due to flooding, the site was previously unused. The designers and planners overcame this challenge through design compatible with the forces of nature instead of against them. Some of the strategies include native plants and flexible paths.

Fig. 6. Bass River Park.

Increasing Sensory Experiences

This principle is based on activities that, along with essential design forms, stimulate the sense and emotions of the inhabitants. This enables the reduction of the gap between the individual and the environment.

Additionally, it also forces individuals to react due to the stimulation of challenge to an individual consciousness through this interactive design. An example of this is visible through David Baker and Fletcher Studio, which created a sustainable urban design proposal for one of the districts of Dallas.

The design was interactive, where people were involved in the city's planting, management, and harvesting process, generating a sensory experience and environmental awareness.

WAYS FOR ACHIEVING SUSTAINABLE CITIES THROUGH AESTHETICS

This section explores two ways of achieving sustainable cities through aesthetics:

Public spaces

A public park that could be utilized by the residents and the employees working around it will be the best-suited option for development for the chosen site. In addition, introducing elements such as a jogging track and central meeting spaces will encourage exercise and provide a shared space for people to spend their leisure time. A play area for children is an essential element that needs to be incorporated into urban design. Furthermore, it will regulate the bioclimate of the surrounding. Proposing a public park will help reduce stress, generate positive emotions, and promote mental well-being among the residents and the employees. It will contribute to a stronger society and a more sustainable environment. The Rotary centennial park is an appropriate example due to its linear structure and planning. Parks and streets an also be incorporated with street furniture and art to enhance the look and give a more appealing look.

Market place

A marketplace possesses qualities such as diversity and inclusiveness. In addition to that, adding functions such as workshop spaces, exhibition areas and design objects will generate interest in people visiting the market area. The structure will generate playfulness, exploration and interaction by creating new interaction pathways among a new and diverse mix of people. The open-air market also creates employment opportunities and will strengthen the economy of the place of location. People living nearby will receive a place where they can move freely, talk freely, and feel in the community. Therefore, colourful central marketplaces is a way to enhance social gathering.

AESTHETIC OF SUSTAINABLE CITIES (FUTURE CITIES)

Throughout history, cities have been at the heart of human development and technological advancements. Although an element of planning can be discerned even in the earliest cities they have often evolved in response to the changing needs and aspirations of their inhabitants. Some cities have survived for millennia, including Rome, Athens, Cairo, Alexandra, Baghdad and Beijing, and are still flourishing. Other, once mighty, cities have disappeared, their ruins being unearthed by present-day archaeologists. A fascinating example, built by the Khmer civilisation, is Ankor Wat in present day Cambodia, which boasted features very relevant to the design of future cities. Notable achievements by the ancient Khmer engineers were the control and distribution of water through a sophisticated canal system irrigating agriculture within the city bounds supplying citizens with ample food. The fundamental problems that the Khmer solved were the prevention of flooding by Monsoon rains, and storing water for the subsequent periods of drought. Despite its success over 8 centuries, AnkorWat collapsed in 1431. Climate change resulting in extended droughts is considered to be a contributory factor, which even the excellence of Khmer engineering was unable to counter.

Tomorrow's cities: chaotic or strategic? Lessons from the immediate past

First and foremost, future cities must serve their citizens, combining increased prosperity for all with desirable life styles. These aims must be achieved without detriment to people who live in other regions; for example they must not export carbon emissions by importing goods manufactured by fossil fuel and feedstock dependent processes or create pollution elsewhere. This is not to say that a city cannot import feedstock or energy intensive goods from outside its borders; rather the energy and material contents of imports must be balanced by those of exports.

To this end, future cities must adopt wide scale utilisation of renewable energy, waste management/minimisation, water harvesting/recycling, landscape/ biodiversity to enhance the natural environment, use of green transport systems, applications of innovative material/construction methods (low/zero carbon buildings) and local food production.

While such aspirations would have been familiar to city designers in antiquity, their modern counterparts can draw upon newer technologies such as integrated smart management control systems based on wireless sensor networks, which by detailed monitoring can turn wasteful cities into sustainable cities. Technologies will need to be tailored to particular geographic, climatic and cultural conditions, but all will have a similar philosophy of turning buildings from passive entities to active, adaptive and adaptable spaces that takes advantage of the surrounding environment for heat, cooling, light and electricity.

A key to achieving low carbon cities is understanding how best to select and integrate various technologies from the many available, to optimise performance for different building types, climates, cultures and socio-economic conditions. A strategic approach will be required to achieve a sustainable city to ensure that it functions efficiently as a whole. But the planning parameters should not be so centralised and rigid that they do not allow different community designs and architecture styles to find expression.

Several impacts of overpopulation in urban areas are already appreciable as summarized in Table 1 with possible mitigation strategies. First of all concerns of food and water security is arising in many cities. As these cities expand, agricultural land is converted into residential and industrial areas. For instance in Conception, a Chilean City with a population of 1734 ,500.000 hectares of wetlands and 1417 hectares of agricultural land and forests were transformed into residential areas between 1975–2000. In Accra, Ghana, it is estimated that 2600 hectares of agricultural land is converted every year where Chinese and Indonesian cities have the similar pattern. In the future agriculture will be challenged to meet the demand of a population that is projected to grow and to urbanize. This indicates that more food will be demanded by a population of net food buyers; and food demand will have to be met by rural and peri-urban areas or by food imports.

Table 1 Impacts of global urbanisation and mitigation strategies

High traffic density	– Efficient public transport
	– Compact city design
High amount of waste	– Recycling
Urban warming	– Increasing green space,
	– Using reflective materials
Increasing Air pollution	– CO2 capture,
	– Filtering exhaust gases,

	– Increasing efficiency of industrial processes /vehicles
Increasing energy consumption/ sinking resources	– Using renewable sources,
	– Achieving low energy buildings,
	– Increasing efficiency of devices/processes
Lack of biodiversity/natural habitat	– Increasing green space,
	– Developing animal/plant protection areas
Sinking water resources	– Water purification
	– Desalination
	– Rainwater harvesting
Rising food demand/poverty	– Vertical farming
	– Artificial food production
	– Greening the deserts
Land shortage for housing	– Constructing multifunctional buildings,
	– Creative architectural designs
Weak Social cohesion	– Improving sociocultural environment
	– Increasing the number of organisations-events that bring people together

On the other hand, in many emerging cities, people are obligated to live in more marginal regions. They have less adaptive capacity, low incomes and no assets. Additionally less legal and financial protection, no insurance, no land tittle... Indeed urban areas are more attractive than rural areas for many people in terms of job opportunities, improved living conditions, multicultural environment and dynamic life. In fact developed countries are already highly urbanized due to the opportunities they have, and the United Nations estimates that the urban populations of Africa, Asia, and Latin America will double over the next 30 years, from 1.9 billion in 2000 to 3.9 billion in 2030. At that point, over 60 % of the world's population will live in cities.

Cities with the growing urban mass will turn to a more resource, land, food and energy demanding consumers and they should be productive to be genuinely sustainable. Unproductive urban areas will probably face with poverty, inequality of individuals, pollution, illnesses and external economical dependency.

Productivity is clearly desirable in emerging cities as it increases competitiveness thereby prosperity and sustainability of any city. More productive cities are able to increase output with the same amounts of resources, generating additional real income that can raise living standards through more affordable goods and services. More specifically, the generated extra income and municipal revenue will enable any city to provide more, better services, including housing, education and health services, social programmes and expanded infrastructure networks to support both productive and leisure activities. Indeed, the productivity of the city is directly influences the citizen well-being and socio-economic status.

Urban productivity is the measure of how efficient a city transforms inputs into outputs. Gross domestic product (GDP) per capita is commonly used as a proxy for urban productivity, with a city's GDP measuring local production of goods and services and the population serving as a proxy for inputs related to human capital. GDP is an important measure of sustainability of developing cities. The report by UN demonstrates that GDP of emerging cities is significantly influenced by the national development level. The importance of national comparative advantage is illustrated by the fact that, while 22 of the top 30 largest urban areas (by population) were located in emerging or developing economies in 2008, only seven emerging economy cities ranked among the top 30 in terms of urban GDP. The group included Mexico City, São Paulo, Buenos Aires, Moscow, Shanghai, Mumbai and Rio de Janeiro, but no Middle Eastern or African cities. The average GDP per capita of these emerging/developing country cities tends to be substantially smaller than that of developed cities.

The recent report by the United-Nations (2014) estimate 3.9 billion urban population in 2014 itself. The urban population will be increase about 2.5 billion, which means by 2050 UN estimates about 6.4 billion population residing in urban areas. Increasing urban population will lead to a significant increase of urban energy consumption and urban emissions. Higher population may also cause an increase in urban density (number of people per unit area). Key strategies are needed for minimizing energy consumption, efficient use of land, sustainable food production and transport.

In a recent study researchers developed a tool for predicting future energy consumption and CO_2 emissions based on electricity, heating and transportation in urban areas. The tool was applied to 3646 urban areas and three projections were used in the analysis for the years 2020 and 2050. In medium and high projection it is assumed that the number of people per unit area (km^2), (urban density), will increase 1 % and 2 %, whereas in low projection no urban density increase was assumed. The results revealed that, for the high projection, CO_2 emissions based on electricity usage will be doubled in 2020 and increase more than four times in 2050. According to the developed tool, heating sector based CO_2 emissions and energy usage will not affected from the urban density and will slightly increase in all cases. On the other hand the increase in urban density may dramatically increase the transportation sourced emissions and energy consumption. While some cities in the past have been planned, at least in part, they have often grown chaotically with minimal strategic design and service provision, especially when under pressure from population growth often driven by migrations from the countryside during periods of agricultural mechanisation and industrialisation. The large industrial cities of 19th century Britain and the favelas of 20th century Rio de Janeiro are obvious examples. Lack of sanitation, overcrowding, disease, child labour and little provision for education resulted in the horrendous living conditions described in the novels of Charles Dickens and the factual reports of investigative journalist Henry Mayhew.

In Britain a century and a half of social reform, legislation and regulation has been required to heal the wounds of past piece-meal urban growth. Even now Britain suffers from a considerable 19th and early 20th century house stock, which whilst impractical to replace, is also technically challenging to refurbish to the modern standards of energy efficiency necessary to reduce national carbon emissions.

To avoid the chaos of the past future cities must be planned. But city wide planning, especially on grand scale, has a legacy of suspicion. Like other European nations after the Second World, Britain, embarked upon a combination of slum clearance in bomb-damaged cities and initiating new cities on greenfield sites, in part inspired by the "concrete" visions of the Swiss-French architect Le Corbusier (Charles-Édouard Jeanneret-Gris), who was active in the first half of the 20th century. Tower blocks and "streets in the sky" constructed using reinforced concrete and industrial engineering-type processes proved to be a disaster, both technically, because of lax quality standards, and socially, because of social isolation and creation of conditions conducive to petty crime. In Britain, 60 years later, these structures are regularly demolished to be replaced by more traditional low rise dwellings built to modern standards.

The key word is "chum" that implies an insensitivity for humans as individuals. On the other hand, influenced by the British planning failures of the 1950s and 60s, heir to the British Throne Prince Charles, is highly critical of "modern architecture" and strongly advocates a return to local vernacular architecture for new communities, a philosophy he has applied to the development of Poundbury in southern England. Charles' ideas have been dismissed as an anti-progressive pastiche of the traditional. But, in some respects, Charles is returning to the earlier vision of British urban planner, Ebenezer Howard the creator of the garden city movement. Letchworth was the first garden city followed by Welwyn Garden City, both built in the early 20th century to the north of London, recognised the importance of an integrated transportation system, but, not surprisingly for the time they were built, did not foresee the rapid growth of automobile ownership. The garden city philosophy was adopted in other countries. In summary, inspired city planning working to a well-defined strategy with a regard for human values can avoid the chaos created by the uncontrolled growth of cities during rapid industrialisation. Engineering has much to offer future city design but it must be tempered by a respect for citizen's aspirations. The designs for future cities must be flexible, responding to evolving technologies and cultural changes. With rapid global urbanisation the challenges are immense; but to learn what works new cities must be built to new designs. We shall need to accept that success will be accompanied by failure, from which we shall, and must, learn.

REVIEW OF THE VISIONS FOR FUTURE CITIES

Innovative visions are needed in emerging cities to reduce the impact on the environment while creating places that increase social cohesion, or accelerating human interaction in education, health and employment to improve the quality of life for an ever greater percentage of our world population. The technological advancements should be fully utilized to realize these visions and goals. For instance, temperature, pollution, water systems, waste management systems, radiation, traffic, air pollution and other components can be monitored through wireless sensor networks for achieving the greatest efficiency.

These systems can help detect leaks and problem areas quickly, potentially saving electricity and other precious resources. In order to save additional resources, cities can consider grassroots initiatives, like farmer's markets and community-supported agriculture. Urban farming is a simple change, since dirt beds can be put nearly anywhere and grow food locally. Organizing community carpools and encouraging people to recycle waste and use reusable bags for shopping can make huge impacts as well. A staggering 75 % of solid waste is recyclable, but steps need to be made to encourage more recycling to happen, as 70 % is still thrown into the trash. Cities can become also more sustainable and attractive by adding open space. Hiking trails, activity centres, and parks can draw people into the city and reduce waste.

Cities are vital to the future global economy. For instance 41 % of the UK's population lives in the country's ten largest urban areas. However, cities are struggling with climate change, changes in population and demographics, congestion and healthcare, and pressure on key resources. In future there will be a large market for innovative technologies/approaches to create efficient, attractive and resilient cities. Recent research has been focused on the development of a data platform for power, heat and cooling usage in cities and individual usage patterns in domestic, commercial and industrial buildings. There is a lack of information in the rapidly changing energy market. Solutions are required to better handling of cost, supply and demand of energy in cities and towns. With macro-level energy data, cities can invest in new innovations, provide more focused geographic support to areas where energy supply is lacking, and gain better decision-making evidence on issues such as targeted building retrofitting and fuel poverty.

Responding to the rapid urban development and challenges, future cities have become a pressing issue due to the impacts of global warming problems. This inevitably requires identifying prioritizing and structuring new design and managerial tools to improve their environmental, urban and fiscal sustainability.

Emerging cities should also develop local and national policies to retain highly qualified individuals. Currently in developing world, the proportion of cities making effort to retain talented and visionary individuals is alarmingly low. Asia could count as an exception where half of the cities are putting effort to retain talent. In China, Chongqing has developed an ambitious training programme to support the transition of rural migrants from manual-based to skill-based types of work; by 2009, nearly one-third of migrants had benefited from the scheme. Dubai is also promoting education especially in the fields of engineering and information technologies.

Some cities in developing countries have embraced the model of world-class innovation clusters, such as California's Silicon Valley or Boston's Route, to become 'high-tech hubs'. Those that have met with success in this endeavour, such as India's Bangalore, owe it to the same basic factors: the presence of top-quality academic and research institutions as well as substantial public and corporate investment.

However, low infrastructure development rate and unbalanced distribution of benefits of growth across all the population are signalling threat for these regions. Quality of life is rapidly emerging as a major asset in any efforts to attract and retain creative minds and businesses. It is not surprising that Toronto, San Francisco or Stockholm are regularly ranked among the top performing cities in the world, since they are found as performing particularly well in a wide range of both economic and quality of life indicators including crime, green areas, air quality and life satisfaction. Except more developed nations, Singapore, with a similar balance of quality of life attributes, also ranks among the top world cities and the highest among developing countries.

Inspiring from the above given successful examples, each city should develop its own strategic future vision for realizing the basic concepts, with the aim of maximizing an integrated total of environmental, social and economic values. When setting out the future vision, both a backcasting approach of looking back from a desirable future to the present and a forecasting approach of looking forward from the present to the future are essential to enhance feasibility. Moreover, it is important to set the vision in a way that fully embodies each city's diverse and unique features that arise from its natural and social characteristics.

Each city is required to tackle the challenges of the environment and aging society, and is further encouraged to take on additional challenges in areas that can enhance their originality and comparative advantages in cooperation with other cities in the same nation and abroad. It will be important to gather worldwide wisdom by absorbing information on other cites' successes from all over the world, as this will help integrate a variety of efforts in different fields and realize synergistic effects. By accumulating successes, cities are expected to break away from subsidies and acquire self-financing independence, establishing financially and socially autonomous models. The European "Smart Cities & Communities Initiative" of the Strategic Energy Technology Plan (SET-Plan) promotes 40 % reduction of greenhouse gases in the urban environment by 2020, which could be achieve with sustainable and efficient production, conversion and use of energy. Yet the domestic sector will increasingly become the leading energy sector as more people around the world aspire to higher living standards, which will drive the demand for air conditioning and electric power. Zero energy buildings (ZEB)/Zero carbon buildings (ZCB), therefore, expected to have a vital role to achieve sustainable and smart cities. Kylili and Fokaides define ZEBs as buildings that have zero carbon emissions on an annual basis.

Various designs for future cities have been mooted, some more adventurous than others. Some are actually being built. All aspire to being carbon neutral and sustainable, exploiting the latest technologies for construction, renewable energy, recycling and transportation.

Recently the British Government has announced plans for new garden cities in the UK which emphasised the development of new communities adapted to local needs. The aspirational "wish list" harks back to Howard and, although arguably obvious, it does express what is expected of a future British garden city:

- Strong vision, leadership and community engagement
- Land value capture for the benefit of the community
- Community ownership of land and long-term stewardship of assets
- Mixed-tenure homes and housing types that are affordable for ordinary people
- A strong local jobs market in the Garden City itself, with a variety of employment opportunities within easy commuting distance of homes
- Beautifully and imaginatively designed homes with gardens, combining the very best of town and country living to create healthy homes in vibrant communities
- Generous green space linked to the wider natural environment, including a surrounding belt of countryside to prevent sprawl, well connected, biodiverse public parks, and a mix of public and private networks of well-managed, high quality gardens, tree-lined streets and open spaces
- Opportunities for residents to grow their own food, including generous allotments

- Strong local cultural, recreational and shopping facilities in walkable neighbourhoods
- Integrated and accessible low-carbon transport systems – with a series of settlements linked by rapid transport providing a full range of employment opportunities

Garden cities built along these lines will largely exploit existing technologies, an approach already adopted elsewhere. The Zero Carbon Building (ZCB) in Hong Kong, is located at the heart of Kowloon Bay, the upcoming vibrant premier business district. Covering a total area of 14,700 m^2 comprising a -3storey Zero Carbon Building and a landscape area it both showcases state-of-the art eco-building design and technologies to the construction industry locally and internationally and raises community awareness of low carbon living in Hong Kong.

To achieve zero carbon emissions, ZCB adopts an integrated design where the ZCB building and its surrounding woodland must be seen as a single entity. Nevertheless, in addition to the sustainability of future's construction we should also consider the "visuality" and "functionality" of buildings. An intriguing example, Hong Kong Polytechnic University's 15-storey Jockey Club Innovation Tower competed last year. It prospers the diversity, expresses the dynamism and creativity of university life with creating a fascinating turban area. While the tower provides multifunctional usage and is visually attractive, its unique geometry covers less land space than its contenders. The building is a showcase of future high rise construction. Future cities could evolve by progressively adding more buildings following the same principles, each designed for its intended function, residential, offices etc.

The principles incorporated into Hong Kong's ZCB and Jockey Club Innovation Tower can also be seen in Swedish developments. Malmö, Sweden's largest city has undergone economic changes replacing its tradition heavy industry with small and medium size companies. The buildings have a number of measures for ecological sustainability using a combination of wind, geothermal and solar energy. The original biodiversity of the local area has been maintained and especial attention has been applied to rainwater collection and sewage treatment.

Hammarby Sjöstad (Hammarby Lake City) is a new district in Stockholm built on a previously industrial and harbor area. Hammarby is meant to provide 10,000 apartments for 25 000 inhabitants and occupies 200 hectares of land, close to the city centre. The required environmental impact of the project was limited to no more than half that of the best projects built at the end of the 1990s; in the long term, the energy demand should not exceed 60 kWh/m^2 per year of which not more than 20 kWh/m^2 per year should be electric energy. As with the Malmo development, energy, waste and water systems have been designed for sustainability. A similar development, Beddington Zero Energy Development (BedZED) in London, was completed in 2002 comprises 82 affordable dwellings and commercial site (offices, workspaces) spread on approximately 2500 m^2. The project is a conspicuous example of urban development as it addresses many challenges such as combining workspace with housing, matching with dense urban population, achieving zero carbon standards and increasing comfort level.

Japan is also actively developing sustainable "eco" cities, of which a particularly interesting example is the Kitakyushu Eco-Town project. Like the Swedish examples, its development is a response to the decline in highly polluting heavy industry, which contaminated the local, land, sea and air in the 1960s. The target is to reverse this environmental damage by creating a sustainable community through a partnership of the government, commercial organisations and citizens. A key aspect is local recycling of discarded items from bottles to bicycles. Furthermore, all Eco-Town companies must allow their facilities to be inspected by citizens in order to eliminate public distrust and anxiety concerning potential pollution.

The developments described above are based essentially on established technologies following principles that can be applied readily elsewhere to achieve urban sustainability in the near future. They are targeted at relatively modest sized communities typically adjacent or within existing conurbations.

In parallel with these projects, far more ambitious, schemes have been initiated that are creating completely new sustainable cities on virgin ground, especially in states with strong central, governments and with considerable national wealth earned from the sale of fossil fuels. A good example, of a future community is Masdar City in Abu Dhabi (UAE), a project to create the world's first low carbon/zero waste sustainable city.

Completely powered by renewable energy, and covering an area of more than seven square kilometres, Masdar City will have the capacity to house 40,000 residents, and host a range of businesses and institutions employing 50,000+ people. But, it is intended to be more than just a demonstration of the practicality of using renewable energy technologies. Masdar City will host a vibrant, innovative, community of academics, researchers, start-up companies and financiers – all focused on developing renewable energy and sustainability technologies.

Another interesting project, Silk City in Kuwait, will be completed in 2023 and will include 30 communities grouped into four main districts; Finance city, Leisure city, Ecological City and the Educational - Cultural city. Silk City will become a new urban centre accommodating 750,000 residents in over 170 thousand residential units. This $132 billion project will create a modern and sustainable oasis, providing hundreds of thousands of jobs and investment opportunities within the world's tallest tower "Burj Mubarak al-Kabir" located in Finance City.

King Abdullah Economic City is another representative of the future city concept aiming to have a positive impact on the socio-economic development of Kingdom of Saudi Arabia. The first stage of the city was finished in 2010 and it will be fully completed in 2020. It will consist of several zones enabling industrial, educational, business and residential activities over an area of 173 km^2. Energy/carbon, water, waste, ecology/biodiversity and pollution prevention have been adopted as key parameters in the design of the city. It is also aimed to create up to one million jobs for the youthful population of the country, with where 40% are under 15.

In response to its considerable environmental problems, a result of its recent industrial growth and need to meet the aspirations of its increasingly wealthy population, China has initiated the construction of many cities based on sustainable designs. In contrast to Europe and Japan, China is able to build on green field sites, an example is Tianjin Eco City in China. Although its development has not been without problems it does appear to be growing at a viable pace. The stated intention is to move one hundred million people into new cities in the next decade, especially in the western part of the country.

Azerbaijan is developing Khazar Islands, a sustainable $100bn city in central Asia on the Caspian Sea, which, when complete in 2020 to 2025 will have 1 m inhabitants. Amenities provided in the city will include; cultural centres and university campuses. The prestigious $2bn Azerbaijan Tower, intended to be the world's tallest, presumably trying to outdo Burj Mubarak al-Kabir. A Formula 1 circuit will also be included. All buildings will be capable of withstanding magnitude 9.0 earthquakes.

While the new cities described above are ambitious they are based on existing or emerging technology and, in principle at least, can be completed within the next decade, designs for far more futuristic cities have also been mooted, siting them underground, underwater, floating on the sea or even in the sky.

Arguably the development of the underground city has already started. In London, where real estate is very expensive, wealthy property owners are digging downwards to expand their living space thus avoiding planning regulations. The London Crossrail scheme shows that large underground spaces can be created, but as Harris notes, quoting London's Road Task Force, why not put major roads into new tunnels "...leaving the surface, with its sunlight and trees, for public spaces"? Maybe in localities such as London, where the underlying clay is conducive to excavation, a present day city can evolve into a future city by digging downwards rather growing upwards? An alternative option for London is Sure Architecture's "Endless (Vertical) City" envisages a 55 storey tower designed for London site which will be a self-contained community complete with areas dedicated to parks. Two ramps wind around the exterior essentially providing "vertical" streets since London does not have the space to accommodate further horizontal streets.

With Japan's lack of building land and susceptibility to earthquakes it is perhaps not surprising that a Japanese company, Shimizu Corporation, has proposed building self-sufficient cities under the sea called "Ocean Spirals". A city with typically 5000 inhabitants will be contained within a 500 m diameter water-tight sphere, at or near the ocean surface, and connected by a huge spiral to the ocean floor as much as 4000 m below. Aquaculture would be practised in the surrounding sea to produce food sustainably and fresh water would be obtained by desalination. Shimizu claims the first city, costing £16bn, could be ready by 2030, having taken just 5 years to build...and the price of further cities would be reduced as numbers increased.

In contrast to Shimizu, Architect Vincent Callebaut has designed the "Lilypad" city, capable of accommodating 50,000 people floating on the ocean surface. The city integrates a range of renewable energies (solar, thermal, photovoltaic and wind). Intriguingly, since these floating cities float near a coast or travel around the world following the ocean currents, they would avoid the problems of sea level rise resulting from climate change.

The Venus Project, proposed by US inventor, Jacque Fresco, is another circular city comprising a central dome containing the cybernetic systems that maintain core automated city functions. Fresco goes way beyond developing a sustainable city. He wishes to create an utopian, technological civilisation without money that avoids the ills of all previous forms of economic and political systems...capitalism, government, fascism, communism, socialism and democracy.

Fresco considers that by creating the ideal environment for humans it will naturally eliminating violence, greed, and the inequalities that presently afflict us. His philosophy seems to be in a tradition that can be traced back to Plato and Thomas More. The ideas espoused are beguiling, but are they achievable? Could they survive in a world where the pursuit of power and wealth is the prime objective of some individuals, whether ostensibly justified by nationalism, religious belief, or political creed? Indeed, to fully buy into the Venus Project requires a strong belief in its philosophy.

Even more fanciful than London's "endless City" and inspired by the form of the lotus flower, is Tsvetan Toshkov's, "'City in the sky' which he claims is a concept embodying an imaginary tranquil oasis above the mega-developed and polluted city, where one can escape from the everyday noise and worries." Although a delightful exercise in creating a utopia away from the strains of modern city life, the engineering stresses within the proposed structure raise questions about its practicality.

Despite the ambitious, indeed grandiose, designs of future cities requiring considerable planning, rapid urban renewal may become vital in response to natural disasters notably earthquakes and hurricanes. While nobody would wish such misfortunate on any city with the human tragedies engendered, the opportunity presented to rebuild a devastated city to both improve its sustainability and to reduce the risk of future disaster cannot be overlooked, not least as an honour to those who have suffered. Two examples are the Wenchuan and Qingchuan districts of Sichuan Province, severely damaged by the 2008 earthquake, which are now in the reconstruction process.

These areas suffered because buildings were not earthquake resistant. Reconstruction has been difficult and a large number of temporary shelters that are neither durable nor thermally comfortable have been built in an attempt to meet the urgent needs of those affected. A research team led by Prof. Zhu Jingxiang of the School of Architecture at The Chinese University of Hong Kong (CUHK) has developed an integrated light-structure system for the reconstruction of New Bud Primary School at Xiasi village in Sichuan's Jiange County.

With the support of the Hong Kong Dragon Culture Charity Fund and the CUHK New Asia Sichuan Redevelopment Fund, the new school was completed and in operation in just two weeks. The building is safe and durable, and the cost of construction is low. It also looks attractive and features good thermal performance and a high energy-saving capacity. Maybe inspired, elegant, but eminently practical designs to rebuild shattered communities rapidly and sustainably will be more important and helpful to humanity than some of the grandiose schemes presently on drawing boards?

BUILDING FUTURE CITIES

Of course, new cities, based on existing modern technology, are already being designed and built. China is responsible for about half of global construction work and will build 400 new cities and towns within the next 20 years. China is also moving rapidly towards implementing a low carbon economy and have recently selected 5 provinces and 8 cities for low carbon demonstration. However, city design and technology must continue to develop, not least because more than 80 % of the world's Global Warming Potential (GWP) is created in cities and by 2050 66 % of the world's population will be urban. Accordingly, advanced construction methods and materials will be needed for sustainability in future's cities.

Robotic/digital design based technologies coupled with 3D printing of prefabricated modules will reduce construction times, minimise energy consumption and eliminate wasted material, all contributing to lower costs. This technology will be an important step for redesign/reconstruction of cities towards sustainability. Basically with the 3D printing, robotic arms with three axis freedom of movement can construct the building, based on the architectural design, which is coded into the controller of the 3D printer. 3D printed buildings provide aesthetics while minimizing the constructional defects which is generally an issue in conventional buildings.

A return to timber as a major building material is especially attractive since each cubic meter of wood can store half tonne of carbon. Can we make buildings that work like trees and cities like forests? New cities will exploit new materials that will deliver greater functionality. For example nano-materials already offer opportunities for advances in sensors and smart polymers. However, it is just as important that future cities are constructed from materials that are completely recyclable and sustainable.

Where virgin feedstock is required it must be taken from renewable sources, which in many instances will be biomass-based. For health and safety reasons manufacturing processes are currently located at distances from major conurbations. In the future processes are required that are low hazard and can be integrated into cities, close to workers homes. The newly emergent disciplines of Green Chemistry and Green Engineering are addressing the development of future manufacturing industry. With buildings being responsible for almost half of all energy consumption and carbon emissions in Europe, new build properties are becoming much more energy efficient and their environmental footprint is being reduced.

The Energy Performance of Buildings Directive, EPBD (EU, 2010) requires that all new buildings shall be "nearly zero energy buildings" (nZEB) by the end of 2020. EPBD is not limited to new buildings, but also covers retrofit of existing buildings because these constitute the majority. Accordingly, building materials are in the spotlight as they have a large influence on building energy consumption, carbon emissions, urban warming and comfort level. Solid wood has been used as a building material for thousands of years, appreciated for being a lightweight, easy reusable and naturally regrown resource (See: Fig. 7). Today, "wooden construction" is innovative and on the rise: timber is again regarded as an ideal green building material.

Recently a 30 storey tower, has been designed by Michael Green for Vancouver, Canada. Once built, it will be the tallest wooden construction, overtaking its competitors Forte Building, Melbourne and Stadthaus, London.

Feeding future cities

According to WHO 50 percent of the 7 bn global population is currently living in cities requiring a land area for farming equivalent to half of South America to produce their food. In the next 40 years there will be 3 billion more people to feed implying 50 % more food production. Since 80 % of the world's population is predicted to be living in cities by 2050, seemingly generating a conflict between using land for agriculture and for cities if the extra food production is obtained via traditional agriculture. But the problem of feeding the inhabitants in future cities may be less severe than we imagine. Historically, some cities at least integrated agriculture into their structure... Ankor Wat and Tenochtitian were mentioned above.

During WW1 and WW2 the gardens and spare ground within British and German cities were turned over to the growing vegetables. Even Einstein cultivated an allotment in WW1, although he was reprimanded for it being untidy. Even the USA increased its food production in WW2 by promoting victory gardens. With the collapse of the Soviet Union in 1990, Cuba lost its supplies of fertilizers and agrichemicals precipitating a crisis in food production.

Fig. 7. Benefits of using wood as building construction material

To survive, Cubans turned to intensive urban agriculture to augment their food supplies, an activity which continues to this day. Ironically, when people are restricted to a diet of smaller amounts of freshly grown local food less in quantity than previously, their general level of health improves, an effect clearly evident in both 1940s Britain and 1990s Cuba. In a recent paper Thebo et al. suggest, based on satellite data, that urban agriculture already contributes significantly to the global food supply since an area within 20 km of cities equivalent to the 28 EU states combined is already being used for agricultural activities.

Researchers suggest the potential for vegetable growing within urban areas would require roughly one third of the total global urban area to meet the global vegetable consumption of urban dwellers. But the urban area available and suitable for urban agriculture varies considerably depending upon the nature of the agriculture performed. They reluctantly conclude that the space required is regrettably the highest where need is greatest, i.e., in more food insecure countries. They note that smaller urban areas offer the most potential as regards physical space.

In the developed world urban food growing is becoming popular perhaps for three reasons: firstly by the middle classes the appreciation that urban food cultivation can re-establish the link between food production and consumption, especially for children, encouraging them to adopt a more healthy diet; to supply free, fresh food for those in poverty and perhaps already relying upon food banks; and ironically for high end restaurants. One example of such community organisations world-wide is York Edible in the city of York, UK.

To reduce their environmental impact future urban dwellers will increasingly grow food within, or at least in the immediate hinterlands, of their cities to avoid the CO_2 emissions associated with food transportation especially over transcontinental distances. It is estimated that each 1 Calorie of consumed food uses currently 10 Calories of oil. But where ground is at a premium, food production might be integrated into future cities by 'Vertical Farming', i.e. multi-tier city farms in the form of glass protected skyscrapers or high rise towers that grow the maximum amount of on a minimum land area.

Although one dedicated vertical farm could feed up to 50,000 people, it is still likely that it will be beneficial for all buildings in future to have space reserved for food production. With the recent developments in photovoltaic (PV) technology it will be also possible to design vertical farms self-sufficient and completely sustainable. The primary energy consumption of vertical farms is for lighting (creating mimic sunlight) and water pumping for irrigation. According to the study results, for the vertical farms with a floor area less than 500 m^2, the available space on roof/façade is enough to install required the required number of PV panels. But once the floor area exceeds 500 m^2 the space on roof/façade is insufficient.

The most ambitious schemes for vertical farms will take a long time to realise, if ever. But some more modest examples already exist, for example in Singapore, Sky Greens has constructed a four storey building using traditional growing systems comprising soil based potted plants on a series of conveyor belts which migrate the plants near the windows maybe once or twice an hour so that every plant gets same amount of sunlight during the day.

The technology increases food production by a factor of ten compared to that of traditional farming on an equal land area. Other vertical farms have been built in Korea, Japan, the USA and Sweden. Singapore, one of the most densely populated countries, is considering a futuristic "floating vertical farm" designed by Forward Thinking Lab of Barcelona. The system basically consists of looping towers that could float in local harbours, providing new space for year-round crops. The concept is inspired in part by floating fish farms that have been in use locally since the 1930s.

The flip side of producing and consuming food is that it creates human waste that must be treated to avoid pollution. Although human faeces and urine have been used historically as fertiliser thus creating an "eco-cycle" (Fig. 8), it has long been discouraged in developed countries because of food contamination by pathogens. This especially applies to uncooked foods such as salads. Progressive build-up of toxic, heavy metals in the soil and thus plants is also a long term problem. But merely treating sewage and discharging the resulting effluent to rivers or the sea loses valuable nutrients, notably phosphorus, and also nitrogen and potassium, which have to be replaced from unsustainable sources. Vancouver based, Ostara has developed the "Pearl Process®" that recovers phosphorus and nitrogen plus magnesium from waste water to produce a slow release fertiliser called "Crystal Green®", which has a low heavy metal content. The first European plant has recently been installed in Slough UK to treat water from a local industrial estate. Although Crystal Green is presently sold for conventional agriculture, technology of this type will be essential for sustainable urban agriculture. Energy input, required to operate the process, can potentially be obtained from renewable sources, especially solar.

TRANSPORT FOR FUTURE CITIES

To overcome rising traffic problems, cities should be compactly structured with improved accessibility, and have a well-designed transport network.

In future cities effective transportation will play a key role. For the health and well-being of citizens walking and cycling between their homes, workplaces, shops and other locations is already being encouraged. Undoubtedly these self-propelled systems will be integrated into future cities, avoiding the modern day perils of mixing pedestrians, cyclists and powered vehicles. For distances and occasions where self-propelled travel is impractical then future cities will need to strike a balance between mass public transport (buses and trams), individually-hired vehicles (taxis and rental cars) and individually owned vehicles. Various technologies already in development will impinge upon the choices made, such as self-driving vehicles, electric vehicles and Aero-Mobil.

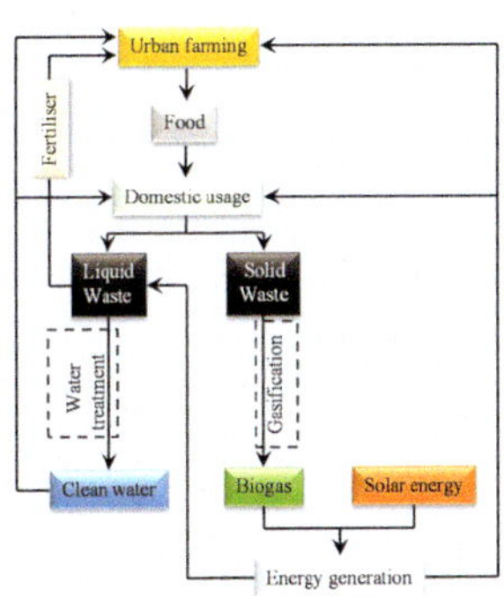

Fig. 8 Eco-Cycle describing the recycling of solid waste and sewage for food, water and energy production for sustainability of future cities.

Aero-Mobil is a flying car that integrates existing infrastructure used for automobiles and planes. As a car it can fit into any standard parking space, uses regular gasoline, and can be used in road traffic just like any other car. As a plane it can use any airport in the world, but can also take off and land using any grass strip or paved surface just a few hundred meters long. It is now finalised and has been in regular flight-testing program in real flight conditions since October 2014. The Aero-Mobil is built from advanced composite material includes its body shell, wings, and wheels. According to company authorities the final product will include all the standard avionics and, an autopilot plus an advanced parachute deployment system.

Undoubtedly safety will be improved over current transport systems, but a serious debate is required concerning citizens' freedom to own their vehicles against potentially more energy efficient mass transport. The driverless, for-hire vehicle summoned by an app may find increasing acceptance. At present road vehicles are dual purpose, being used for both modes but in the future a distinction may be drawn between *intra*-city transport and *inter*-city transport. For long distances the speeds of trains may increase so journey times rival those of intercity aircraft, but operate with superior energy efficiency, for example maglev trains exploiting superconductivity and the "Aero-train" that is part train and part aircraft.

Traffic and transportation are growing problems for all cities. In Europe, people are wasting 10 to 60 hours in traffic jams each year, while their vehicles are contributing significantly to global warming by emitting carbon dioxide emissions and to air pollution by emitting nitrogen oxides and carbon particles. Therefore traffic management and monitoring systems are currently already being applied in many large cities and people are strongly encouraged to use public transport instead of personal vehicles. Even though some steps have been taken more radical, innovative will be needed. Perhaps "futuristic" solutions can no longer be viewed as engineers' fantasies but essential to avoid increasing urban transport problems.

Chinese engineers have allowed a Hongqi Q3 car to navigated itself through traffic to a destination 286 km away guided by cameras and sensors. It is clear that computers can be safer drivers than human beings. They can react quicker, they can look in all directions at once and they don't get distracted. They won't speed or cut people up. Principally, self-driving cars will be connected with a wireless network similar to the internet or telephone network and all cars will be travelling on major roads under control of satellite and roadside control systems. A traffic jam will be predicted before it even happens by using roadside sensors, GPS and other advanced software.

An alternate innovative design for future transport is the Aero-train which is partly train and partly aircraft. The vehicle is designed with wings and flies on an air cushion along a concrete track using wing and ground effects. This minimises the drag effect allowing the aero-train to consume less energy whilst reaching higher speeds than the conventional trains.

Another imaginative idea, first proposed by Robert M. Salter in the 1970s, is the evacuated tube transport (ETT) where a vehicle occurs in a vacuum to eliminate air resistance and friction. The ETT system envisions superconducting maglev trains operating at speeds of up to 6,500 km/h (4,039 mph) on international trips - that's New York to Beijing in two hours! Although the proponents say that ETT could be 50 times more efficient than electric cars or trains it is only a concept that is the subject of ongoing research. But the achievement of ETT would revolutionise future long distance transportation.

ADMINISTERING FUTURE CITIES

According to the UN report 70 % of the world's population will live in urban areas by 2050 with the number of cities expected to exceed 2000 by 2030, compared to 1551 in 2010 Whilst there are 43 'large cities' with populations between 5 and 10 million in 2014, there are expected to be 63 by 2030. The UN estimates that there will be more than 40 mega-cities worldwide by 2030, each with a population of at least 10 million, compared to 28 today. It is projected that Delhi, Shanghai and Tokyo will each have more than 30 million people by 2030, and will be the world's largest urban agglomerations. This massive global growth of urban areas will requires developments in administrative systems to ensure that technological advances described in previous sections truly deliver improved living conditions for all urban dwellers.

Although the well-established scientific basis for global warming is well established, its full impact appears to be several decades in the future, action is required now to ameliorate its effects by identifying, prioritizing, and structuring new design and managerial tools to improve urban environmental and fiscal sustainability. Despite the vociferous assertions of those denying man-made global warming, the ill effects of "local warming", known as the "urban heat island effect" (UHIE), are already manifest in large cities, especially in the tropics and serves as early warning of what is likely to happen worldwide as global warming becomes more pronounced. UHIE refers to the urban temperature being higher than that of the surrounding countryside, 3 K being not a typical, a combination of solar heat absorption by buildings, roads etc., heat emitted by vehicles, and the conversion of electrical to thermal energy, for example by air conditioners.

The UHIE does not just cause discomfort for urban inhabitants, it is also a killer. Various studies of temperature related excess mortality using historical data have shown that during heat waves above a threshold temperature deaths increase significantly with each further degree rise. Not surprisingly, the young, old and those with serious medical conditions, a most vulnerable.

If a city experiences a heat wave of lasting longer than 5 days the UHIE results in excess mortality that raises steeply with every extra degree of excess temperature during night time; the correlation is weaker with the daytime temperature. Not surprisingly, the young, the old and those with existing medical conditions are most at risk.

Mitigation technologies such as increasing green urban space and biodiversity, use of reflective materials, decrease of anthropogenic heat levels and use of low temperature natural sinks (such as ground or water bodies) aiming to counter the impact of the phenomenon are rapidly being developed and applied in real projects. Rehan provided a detailed framework, including several measures that will diminish the accumulation of heat in urban areas and mitigate their UHIE by a set of planning actions as a strategy to cool the cities. Richer urbanites can in principal, offset the effects of the UHIE merely by turning up their air conditioning or installing more powerful units. But such would be socially reprehensible since it would increase urban temperatures further to the disadvantage of poor who cannot afford a/c or the power to run it. Nevertheless, to protect the vulnerable it may be necessary to build air conditioned refuges where they can be sent when local temperatures are high.

Administrators are already aware of the need to incorporating UHIE mitigation as cities are further developed and is required in temperate regions as well as the topics. For example Public Health England has recently published an excellent guide the adverse effects of high temperatures and methods to combat them, both short and long term.

Although excess mortality caused by the UHIE is occurring now, it will be exacerbated by the rise in global temperatures projected by the Intergovernmental Panel on Climate Change (IPCC). Those who argue strongly that man-made global is a myth and therefore nothing needs to be done to mitigate it, must face the consequence if the world follows their lead and they are wrong people, especially the poor and vulnerable will die. The adverse effect of increasing temperatures is based on sound research and historical data. It is not a theory derived from a computer model. Planning and building "cool" cities must be a primary objective of administrations now to minimise deaths.

In large cities excess mortality from attributable to high temperatures is exacerbated by air pollution, notably NO_x from internal combustion engines that ozone produced by the sunlight-induced reaction of oxygen with unburnt hydrocarbons. Indeed, separating excess urban mortality arising from pollution and high temperatures is problematical. Fortunately mitigations for both tend to be the same...for example by reducing fossil fuel combustion to supply electric power by increasing renewables, reducing the number of internal combustion engine vehicles coupled with the provision of better public transport and "greening" cities by planting more vegetation capable of both cooling the air and absorbing airborne pollutants.

Originally, "green" infrastructure was identified with parkland, forests, wetlands, greenbelts, or floodways in and around cities that provided improved quality of life or "ecosystem services" such as water filtration and flood control. Now, green infrastructure is more often related to environmental or sustainability goals that cities are trying to achieve through a mix of natural approaches. Examples of "green" infrastructure and technological practices include green, blue, and white roofs (See: Fig. 9); hard and soft permeable surfaces; green alleys and streets; urban forestry; green open spaces such as parks and wetlands; and adapting buildings to better cope with floods and coastal storm surges.

Fig. 9. A view of an urban green roof in Chicago, USA

The climate adaptation benefits of green infrastructure are generally related to its ability to moderate the expected increases in extreme precipitation or temperature. Benefits include better management of storm-water runoff, lowering incidents of combined storm and sewer overflows (CSOs), water capture and conservation, flood prevention, accommodation of natural hazards (e.g., relocating out of floodplains), reduced ambient temperatures and urban heat island (UHI) effects, and defense against sea level rise (with potential of storm-surge protection measures). The U.S. Environmental Protection Agency (EPA) has also identified green infrastructure as a contributor to improving human health and air quality, lowering energy demand, reducing capital cost savings, increasing carbon storage, expanding wildlife habitat and recreational space, and even increasing land-values by up to 30%.

Green infrastructure (GI) in urban areas can ameliorate the warming effects of climate change and the UHIE. In a study performed by researchers it is found that that increasing the current area of GI in Greater Manchester, UK by 10 % (in areas with little or no green cover) would result in a cooling by up to 2.5 °C under the high emissions scenarios based on UKCP02 predictions. The potential benefits of increased green infrastructure/green space on reducing UHIE are presented below:

- Trees and shrubs provide protection from both heat and UV radiation by direct shading (both of buildings and outdoor spaces).
- Evapotranspiration reduces the temperature in the area around vegetation by converting solar radiation to latent heat.
- Lower temperatures caused by both evapotranspiration and direct shading lead to a reduction in the amount of heat absorbed (and therefore emitted) by low albedo man-made urban surfaces.

These changes can only be achieved by a city administration that has both the appropriate legal power and the will to serve all its city's inhabitants, the poor as well as the rich.

Much of what has been discussed above has ignored the differing sizes of conurbations.

Cities face different impacts, depending upon their sizes and levels of development. Small cities of upper income nations are facing with population decline as a result of the migration to larger cities for better job opportunities and higher life standards. Diminishing manpower makes it difficult for small cities to compete globally in terms of economy and productivity. On the other hand large cities in developed world are facing with the impacts of aging infrastructure and population. Increasing population creates inequality and social cohesion inside the cities while job opportunities become more competitive.

In contrast to developed nations, small cities in developing countries are faced with the impacts of weak economies and weak urban governance. Due to their inadequate infrastructure and buildings, such cities lack the resilience to survive natural disasters such as earthquake or flood is very low. This Survival is threatened and in many cases many people are forced to vacate their homes . Even large cities in the developing world with inadequate/insufficient transport as well as poor housing stock, are threatened by demographic challenges resulting in social and economical inequality. Environmental pollution is probably the most significant problem facing these cities, a result of the rapid industrialisation. But the latter potentially creates the wealth that can enable developing world cities to overcome their growing pains provided it is harnessed for benefit of all and is not siphoned off by corruption.

In many cities of China and India, environmental pollution, particularly $P.M_{2.5}$ (e.g. oil smoke, fly ash, cement dust) level is sharply rising and threatening the human health as $P.M_{2.5}$ particles have the ability to penetrate deep into the lungs and can cause severe health problems.

The severe impacts of rapid urbanisation can be ameliorated by applying creative design to infrastructure. Ecosystems like multifunctional units will provide several uses rather than a single functionality thereby saving energy, time and cost. For instance garden plots can serve as water management system while providing food for citizens. Similarly multifunctional buildings could save time for people while allowing efficient use of land.

Significant advances in computer simulation provided tools that enable us to evaluate current conditions and requirements thus modelling future scenarios. This phenomenon will have increasing importance in future cities to monitor existing conditions for efficient use of capital and natural resources or controlling traffic flow through wireless sensor networks. In addition it will allow modifying energy usage or household waste of urban dwellings with real time feedback. South Korea has already put this technology into practice in city of Songdo, where traffic, waste and energy usage are monitored. Similarly in Rio de Janeiro there is a high-tech centre where public safety responses to natural disasters or building collapses are quickly identified. The recent earthquake in Nepal demonstrated that, this kind of technological centre could save many lives with timely intervention during disasters.

Technically, highly automated management systems are very attractive, but they have potential downsides. The amassing of large amounts of data about individuals' daily lives is already creating grave concern, both via the internet and CCTV. Increasing data collection could offer the potential for city authorities to exercise greater control over citizen's lives. Technology must be tempered by democratic safeguards if individual liberties are not to be infringed. The vulnerability of a highly networked city to a physical or a cyber- attack on data centres must be minimised. A fascinating concept currently being pursued is the adaption of the "block-chain" algorithm underlying "bitcoin" to the administration of organisations to increase transparency and to minimise corruption.

For example recently "Ethereum" launched is a block-chain platform that allows secure systems to be developed with transactions permanently and transparently recorded.

Large corporations already experimenting Ethereum, include UBS and Barclays. "BoardRoom" is a block-chain dapp (distributed app) founded by Dobson who claims that it can be used to run large organisations by "collaborative decision making".

These developments could potentially be just important to the operation of modern cities as the new engineering technologies.

SOCIO-ECONOMIC DEVELOPMENT AND PROSPERITY OF FUTURE CITIES

Emerging cities should be where human beings find satisfaction of basic needs and essential public goods. Where various products can be found in sufficiency and their utility enjoyed. Future cities should also be the habitats where ambitions, aspirations and other immaterial aspects of life are realized, providing contentment and happiness and increasing the prospects of individual and collective well-being. However in many developing cities, prosperity is absent or restricted to some groups or only enjoyed in some parts of the city.

Low purchase power contrarily increasing expenses could socioeconomically pressurize individuals and minimize their social subsistence. This situation will turn citizens from productive and creative individuals to the ones just trying to survive. "Liveable" cities should support affordable living choices, provide citizens options to have a social status and life conditions independent than their income. Cities also should be compact structured with improved accessibility, they should include natural habitats allowing biodiversity and socialisation of individuals and should have a well-designed transport network which will eliminate the need for private vehicles to overcome the rising traffic problem in growing cities. Besides they should offer a profusion of public goods, develop actions/policies for a sustainable use and more importantly should enable equitable access to 'commons' in order to ensure well-being of citizens.

The future urban configurations should concentrate on efficient use of resources and opportunities that could help to achieve prosperity and citizen well-being in five dimensions as defined below and illustrated in Fig. 10;

- Contribute to economic growth through productivity, generating the income and employment that afford adequate living standards for the whole population.
- **Deploy the infrastructure, physical assets and amenities** – adequate water, sanitation, power supply, road network, information and communications technology etc. – required to sustain both the population and the economy.
- **Provide the social services** – education, health, recreation, safety and security etc. – required for improved living standards, enabling the population to maximize individual potential and lead fulfilling lives.
- Minimize poverty, inequalities and segments of the population live in abject poverty and deprivation.
- Protect the environment and preserve the natural assets for the sake of sustainable urbanization.

The past few decades have witnessed a notable surge in economic growth, but one which has been accompanied by an equally daunting degree of inequity under various forms, with wider income gaps and deepening poverty in many cities across the world. Economic inequality is seriously detrimental to the equitable distribution among individuals of opportunities to pursue a life of their choosing and be spared from extreme deprivation in outcomes. According to recent reports, income gaps between rich and poor are expanding in both developed and developing countries.

In OECD countries, inequalities are as steep as they have been for over 30 years. In advanced economies, the average income of the richest 10% of the population is about nine times higher than that of the poorest 10%. In Europe's Nordic countries, the average is a multiple of six but growing, compared with multiples of 10 in Italy, Korea and the United Kingdom, and up to 14 in Israel, Turkey and the United States.

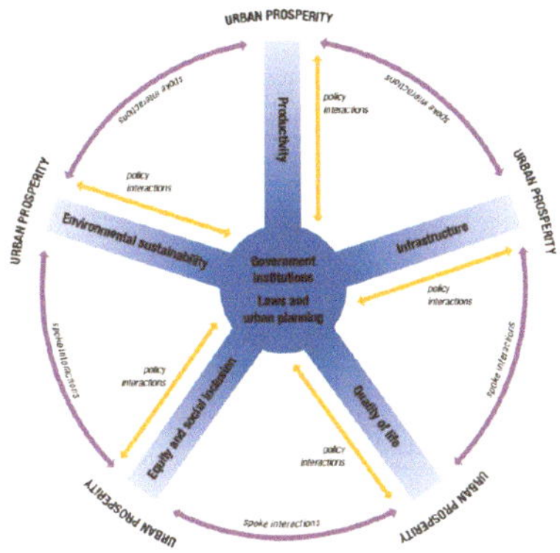

Fig. 10. Wheel of urban prosperity

Cities must realize that equity has a significant impact on socio-economic performance, since the greater the degree of equity, the greater the chances of a wider, more efficient use of available resources, including skills and creative talent Urban prosperity thrives on equity, which involves reduction in barriers on individual/ collective potential, expansion of opportunities, and strengthening of human agency and civic engagement. Cities generate wealth, but the problem is the unequal distribution of it. Despite considerable increases in productivity (e.g. GDP per capita) along with reductions in extreme poverty, inequality as a whole is growing in most parts of the world – a process that undermines urban life quality. In many cities, the population and local experts concur that inequalities are becoming steeper which could be a threat for emerging cities in terms of their sustainability and well-being of citizens.

CONCLUSIONS

The "Future Cities" topic employs a multidisciplinary approach to address the urban development challenges facing emerging cities. This can integrate environmental technologies, comprehensive urban development, fiscal sustainability and good governance, to provide emerging cities with a set of tools in order to improve the quality of life globally.

New-born babies in developed countries are projected to have a life expectancy of 80+ years, with the majority living in cities, increasing yet further the demand for energy, water, food, housing and other services. However, cities are struggling with climate change, changes in population and demographics, congestion, healthcare, and pressure on key resources. In the future innovative technologies/approaches will create considerable market opportunities to transform existing conurbations into the efficient, attractive and resilient cities of the future.

Nevertheless, simply applying innovative technologies alone will not guarantee the combination of sustainability and acceptable living standards for future cities... good governance and management will also play a pivotal role. This can only be provided by utilizing technological advancements optimally whilst also developing short and long term management, organization and development strategies to realize the desired objectives.

CHAPTER

AESTHETICS: THE FINE PRINT 6

INTRODUCTION

Global trends show that annual rural–urban migration has resulted in increased population of most towns and cities especially in developing countries of the world. According to researchers, SSA has experienced the highest urban growth since the 1990s at an average rate of 3.5% per year. Major cities in SSA such as Lagos (Nigeria), Libreville (Gabon) and Accra (Ghana) are characterised by high rates of urbanisation which have resulted in upsurges of unplanned urban settlements. Ghana, a developing country in West Africa, is no exception to this occurrence as its urban population increased from 43.8% in 2000 to 50.9% in 2010.

According to researchers, high rate of urbanisation has been the number one cause of urban sprawl which results in decreasing valuable green spaces within and surrounding expanding cities. In Ghana, for instance, deforestation of surrounding urban forests, as well as, the depletion of existing green spaces within cities are some of the major environmental problems affecting urban ecosystems. According to researchers, Ghana lost an estimated 2.51 million hectares (33.70%) of its forest cover between 1990 and 2010. Interestingly, during that same period, Ghanaians living in urbanised areas increased from about 5 million in 1990 to almost 14 million in 2010.

Additionally, many urban development projects in Ghana and SSA are also characterised by little or no proper attention to green infrastructure development, as well as the ecological and aesthetic components of urban landscape planning. The principles of eco-aesthetics are seemingly deficient in public parks and gardens in Ghana which have contributed to their unsustainability over the years. Many of the public parks created for social recreation in Ghana such as the Tema Children's Park, Efua Sutherland Children's Park in Accra and the Kumasi Children's Park are in deteriorating physical conditions and have thus been abandoned by the public. However, the cultural importance of public parks for social recreation, as well as their ecological benefits to the environment (biodiversity and micro-climate adaptation) cannot be overemphasised. Urban green infrastructure networks of parks, forestlands, gardens, green corridors and waterways also help enhance natural processes of air, water and temperature, reduce stress, and promote the mental well-being of urban residents. Eco-aesthetics (Ecological aesthetics) as a concept promotes the planning of cities and communities in accordance with nature (green infrastructure). The principles of eco-aesthetics also seek to reconnect urban residents to their natural environment by developing attractive green cities that promote environmental sustainability.

Most importantly, the need to plan cities in accordance with nature or green infrastructure has risen in recent years due to the evident changes in global climatic patterns and growing occurrences of natural disasters affecting both developed and developing countries. According to the United Nations Environment Programme (UNEP), SSA is the most vulnerable region to the impacts of climate change and natural disasters such as floods and droughts which occur recurrently. Ghanaian cities, for example, have experienced several incidences of severe droughts, fire outbreak and floods in the past decade which have resulted in the loss of human lives and destruction of properties. Since most natural disasters cannot be completely prevented, there is a global call to build more resilient cites with capabilities to survive and thrive amidst increasing environmental problems. Like in most advanced countries of the world, building resilient cities in Ghana and SSA will help lessen the vulnerability of urban systems, infrastructure and people to the impacts of climate change and increasing environmental disasters.

While previous works on the benefits of eco-aesthetics have not been directly linked to building resilient cities, this section contends that it presents opportunities for cities, especially in SSA to enhance their urban resilience through the use of sustainable green infrastructure networks based on the principles of eco-aesthetics. Again, going through existing literature, there is an apparent gap in Afrocentric perspectives on eco-aesthetics and thus, this section attempts to help abridge this existing gap by exploring the potential contributions, as well as likely barriers to implementing eco-aesthetics in Ghana; a typical developing country in SSA. The main aim of this section, therefore, is to explore the fundamental theories of eco-aesthetics as developed from the West, and demonstrate its potential contributions to building resilient cities in Ghana and other developing countries in SSA.

Interestingly, and against the often-repeated common phrase, the experience of beauty is rarely only 'in the eye of the beholder' or a matter of purely subjective and illogical opinion. As philosophical study of aesthetics points out, judgement of taste is instead a much more complex phenomenon that unfolds as a result of broader cultural, historical, and intersubjective processes. Further this explicates the connections between everyday actions and aesthetic values since it is of vital importance to take greater responsibility for aesthetic choices and evaluations in the face of the urgency of current sustainability challenges. The aim is to show that a better understanding of aesthetic processes can be used to support tactics to move beyond trend-based consumerism.

AESTHETICS AND THE MANIPULATION OF VALUES

It might seem paradoxical to discuss the concepts of aesthetics and sustainability together. Aesthetics seems to refer only to external features and qualities: to the appearance of objects, people, and places. The very word 'aesthetic', in its everyday usage, is associated with the superficial, visually emphasized layer of the human lifeworld. This focus on the perceptually mediated materiality and physicality of very different types of phenomena is, however, a key to understanding why aesthetics is important in solving many contemporary sustainability challenges.

It has been long debated in philosophical aesthetics whether the subject concerns only the aesthetic or if it is more linked to other values. In the Western tradition, the claim of the disinterestedness of the aesthetic experience has been a central idea since Immanuel Kant explored the topic systematically in his Critique of Judgment. In environmental aesthetics, disinterestedness is used to explain the disassociation of the aesthetic appreciation of an environment 'from the appreciator's particular personal, religious, economic, or utilitarian interests, any of which could impede aesthetic experience'.

Following from this idea, a certain disassociated aesthetic attitude prevails when evaluating phenomena that pique our interest with their perceptually manifesting qualities: one has a certain mindset or stance toward the objects (or landscapes, artefacts, people, etc.) of our appreciation or even when initially acknowledging that they are of aesthetic interest. However, the grave contemporary ecological concerns create moral concerns beyond the scope of disinterestedness since they are radically altering the vantage point to the changing aesthetic qualities of the environment.

It needs to be emphasized that the aesthetic, in contemporary theories, is most often understood so that it does not refer only to aesthetically positive qualities such as beauty, the sublime, picturesque, or cute, but also to aesthetically negative qualities such as ugliness, grotesqueness, or even disgusting features, as long as they wake some level of attention and interest. The limits of aesthetic interest are also debated, especially in the subfield of everyday aesthetics: according to some theorists, the so-called restrictivists, everyday aesthetics concerns only those phenomena that are aesthetically elevated among the more mundane experiences, whereas expansionists claim that the very mundane, subtle, and even barely discernible qualities of everyday objects, activities, and phenomena are also aesthetic and should be studied as such through more nuanced conceptual distinctions.

One way to distinguish between different nuances in aesthetic appreciation is to categorize experiences and ensuing judgements into the 'thin sense' and the 'thick sense' of the aesthetic. This distinction between the surface aesthetic (focusing on physical appearance and formal aesthetic qualities such as colour, shape, and composition) and the deep-seated layer of aesthetic judgement is the key to understanding how aesthetic values are intricately tied in with other values—such as ethical and epistemic or cognitive values—which are important to the formation of ecological understanding. Ecological values are of specific interest in relation to sustainability, but it is important to understand that ethical values contribute more broadly to supporting different scopes of sustainability.

Through behavioural economics, nudging has recently gained widespread interest as a concept that sheds light on the mechanisms of soft manipulation. With the means of careful deliberation in the placement, order, presentation, and arrangement of objects and services, for example, people can be steered toward making better choices for themselves. What is better or desirable is determined by prevailing ideas on what is healthy, ecological, socially beneficial, and so on. These ideas are based on continually updated scientific knowledge in relation to the prevailing ideologies of society.

As a form of soft or libertarian paternalism,1 nudging aims at gently directing people toward making better choices for themselves and the community. Interestingly for aesthetics, many of the examples used are focused on manipulating people's attention and perceptual processes. An often-cited example is how food perceived as healthy is placed at the beginning of the buffet table so that the plates of the hungry buffet-goers will be filled with them first. Developing the concept of nudging takes these ways to manage behaviour with the means of choice architecture further and makes them explicitly available as a method of governance in contemporary societies.

The theory of nudging acknowledges that what one sees, hears, smells, tastes, or feels is what one's attention will more likely be directed toward. Sudden changes in perception draw attention more easily, and human beings might be intrinsically interested in new phenomena. Novelty value has been studied in Western philosophical aesthetics to some extent, most notably in the analysis of the 'charm of novelty' but, recently, it has been of interest in more applied fields such as marketing, consumer, and innovation studies. The same applies to other psychologically explained, aesthetically relevant phenomena, such as the Diderot effect, which is used to manipulate consumer behaviour toward making additional purchases that they did not need in the first place.

THE VALUE OF AESTHETICS TO LANDSCAPE ECOLOGICAL PLANNING

The question of how aesthetics affects our ecology in relation to how it connects people to their environment has been tackled from Eurocentric perspectives by researchers. The word "aesthetics" is a complex philosophical term that is used to imply the nature of beauty, art, taste and with the creation and appreciation of beauty. Philosophers such as Immanuel Kant (1724–1804) and Eaton (1998) in their writings help provide a reasonable understanding of the concepts of aesthetics, taste and beauty. Kant asserts that:

> If he says that canary wine is agreeable he is quite content if someone else corrects his terms and reminds him to say instead: It is agreeable to me"... (because) "Everyone has his own (sense of) taste" (The case of 'beauty' is different from mere 'agreeableness' because) "if he proclaims something to be beautiful, then he requires the same liking from others; he then judges not just for himself but for everyone, and speaks of beauty as if it were a property of things.

Kant uses the metaphor of the wine taster as an example to explain the difference between taste and beauty (aesthetics). To Kant, taste differs from beauty because taste is subjective (individually-preferred), whereas beauty or aesthetics is a generally accepted attribute of things. Researchers, however, defined aesthetics as any object that seeks to arouse feelings of delight. Unlike Kant's notion of beauty, Eaton further expanded the definition of aesthetics by emphasising that the term "aesthetic" remains constant whereas "aesthetic values or qualities" are relative to time and cultural factors. In the field of architecture, first century (AD) Roman architect Vitruvius, in his "The Ten Books on Architecture" emphasised how any good architecture work should encompass the three elements of "firmitas" (firmness), "utilitas" (usefulness) and "venustas" (beauty or aesthetics). Venustas or aesthetics according to Vitruvius provides the added value of human pleasure and appreciation of any architectural building or landscape. That is why most works of architecture are generally perceived as a thing of good taste and successful when people perceive them to be beautiful.

Notably, to advocates of eco-aesthetics, green spaces need not only be perceived as a natural resource to be exploited or harvested, but rather, green areas are self-sufficient bodies that must be capable of being aesthetically admired. Aesthetic elements, for example, have been applied successfully in the design of public green spaces in US and other parts of Europe. Frederick Law Olmsted's 1857 plan for New York Central Park, is a classic example of how beauty or aesthetics were interpreted into landscape design for public appreciation and long-term sustainability. Olmsted used unique water and green features interconnected with "organic" pathways to enhance both ecological and aesthetic values of the New York Central Park.

Aesthetic values of landscapes, however, are not always aligned with ecological values in some green spaces. According researchers, people generally perceive well-manicured and trimmed landscapes as attractive, whereas they perceive natural looking or bio-diverse landscapes as messy and unattractive. Subtle landscape forms such as treeless plains and wetlands, as well as ecologically valuable natural forestlands or wilderness may also be perceived as "scenically unattractive"). The concept of eco-aesthetics, however, does not promote the identification of beautiful landscapes, but rather it promotes the aesthetic appreciation of landscapes that provide ecological benefits to the environment. Eco-aesthetics thus integrates aesthetic design elements such as unity, contrast, symmetry, order, etc., in landscape planning and design in order to help evoke human appreciation of ecological sensibilities.

THE VALUE OF CULTURE IN ECO-AESTHETICS

According to researchers, culture or cultural preferences have always been linked to ecosystems. The nature of ecosystems and their conditions such as biodiversity, native species, climate conditions and natural processes have always influenced or directed human cultural systems around the world especially relating to issues of habitat design, religious practices, heritage values, and social interactions. Ecosystems are also capable of providing people with certain "cultural amenities" such as aesthetic appreciation, recreation, spiritual fulfilment, and intellectual development. In Ghana, for instance, certain cultural values or beliefs are attributed to some ecological landscapes such as sacred groves. Sacred groves like the Malshegu sacred grove near Tamale and the Buabeng-Fiema Monkey Sanctuary in Ghana are typical examples of how particular forests and native or endangered species believed to be sacred to the local inhabitants have resulted in the preservation of some natural landscapes in Ghana.

In relation to aesthetics, culture also affects the way a person or a group of people perceive beauty and this is related to Pierre Bourdieu's concept of habitus and taste. How a particular group of people perceive their landscape to be attractive or not is always influenced by their acquired social and cultural experiences both individually and as a group. This explains why people from particular cultures such as in Europe or the Americas have specific aesthetic preferences and values that vary significantly from people in other cultures such as in Asia and Africa. The value of culture to eco-aesthetics is therefore highly relevant in the Ghanaian or African socio-cultural context where its people have diverse cultural preferences and values that affect the way they might accept, appreciate and protect their environment. Nonetheless, other research and theories in eco-aesthetics suggest that there are specific aesthetic preferences of landscapes generally desired by all people and across all cultural dynamics. These aesthetic preferences such as "presence of order", "balance" and "elementary qualities" are referred to as "evolutionary aesthetics", and are believed to be embedded in our human instincts developed during human evolution.

The concept of "cultural ecology" therefore places emphasis on the relevance to cultural processes such as values, beliefs and social organisation in the design of human-oriented environments. Architects and landscape designers such as Ebenezer Howard (1850–1928), Frank Lloyd Wright (1867–1959), Philip Johnson (1906–2005), Ian Mcharg (1920–2001), Jane Jacobs (1916–2006) and Anne Whiston Spirn are all advocates for ecologically friendly urban environments that utilise aesthetics to help reconnect urban residents to nature. In order to successfully do so, cultural values need to be integrated into the design of human oriented settlements. Researcher proposes the concept of "cues to care" where familiar cultural or native design elements should be integrated into the design of urban landscapes to help promote their sustainability over the long term.

Eco-aesthetics and environmental sustainability

The growing incidences of environmental problems such as urban pollution, natural disasters and depletion of green areas especially in SSA have prompted the need for more sustainable approaches to urban planning and design. According to researchers, urban lands or cityscapes are the receptacles that hold all physical and cultural aspects of people, and thus, they need to be sustained for present use as well as preserved for future generations. Although aesthetics is not ontologically linked with general definitions of sustainability, Researcher suggests that aesthetics forms a significant part of realising "culturally sustainable" urban landscapes. The principles of eco-aesthetics promote the creation of ecologically responsive green spaces that are aesthetically pleasing to users or inhabitants in order to make them sustainable. The role of aesthetics or art in sustainability has been further emphasised by researchers who assert that:

> Sustainable technologies appeal to our rationality, but...it is a small part of what makes us care about what we care about. To affect values, to create desire, to make people care about something, you have to affect people's hearts, and bodies, our unconscious dream lives and our imaginations. This is the work art [aesthetics] can do so well.

According to researchers, adopting the principles of eco-aesthetics will help heighten the awareness and sensitivity levels of people's interactions with their environment and their impacts on it. When people find their environment and natural surroundings attractive, they tend to appreciate it better, care for it and are more likely to sustain it for future generations. In Ghana, for example, environmental issues of urban pollution and poor maintenance practices characterise many of its green spaces such as urban parks, green corridors and open spaces. The pollution of the environment in cities in Ghana consequently results in regular outbreaks of epidemics such as cholera and dysentery especially in the capital city of Accra.

Also, illegal small-scale mining activities in Ghana, Cote D'ivoire, Nigeria and South Africa are destroying several cultural sites and forest reserves in these countries. In Ghana, operations of illegal miners have left several acres of land in mining communities desolate and degraded, and polluted major water bodies such as the Pra, Birim and Densu Rivers in Ghana. These water bodies service major water treatment plants in the country that supply treated water to urban communities; hence, their pollution causes higher water treatment costs and a looming water delivery crisis to urban residents.

Care and protection of the environments in Ghana are dangerously inadequate and pose a threat to public health and well-being which intend lessen the quality of life of the people. Other environmental challenges such as recurrent floods and droughts create fragile ecosystems and environments that will undermine the achievement of the Sustainable Development Goals (SDGs) in Ghana. Urban environments in Ghana therefore require innovative urban development strategies directed towards building urban resilience. The following section thus explores the concepts and principles of a resilient city.

RESILIENT CITIES: UNDERSTANDING CONCEPTS AND PRINCIPLES

The concept of "resilient city" has in recent years become a major subject of academic discourses. Resilience as a term emanates from the Latin word *resi-lire*, which means to "spring back" and was first used in the field of ecology in the 1970s to refer to the ability of a system to maintain stability or recover functionally when exposed to disruption or disturbance. A resilient system would therefore be one that can resist, absorb and recover from the effects of a disturbance or hazard promptly and efficiently.

The concept of resilience is highly relevant to urban planning because the city represents a complex network of interconnected systems that require constant adaptations to changing conditions. The Rockefeller Foundation defines a resilient city as the capacity of people, systems and infrastructure within a city to survive, adapt, and grow in the midst of increasing "chronic stresses" and "acute shocks". Chronic stresses are defined as persistent pressures that weaken the fabric of a city on a daily or cyclical basis such as high unemployment, inefficient public transportation systems and droughts (that cause food and water shortages). Acute shocks on the other hand, are sudden, sharp events that threaten a city such as earthquakes, floods, disease outbreaks or terrorist attacks.

Assessing and improving a city's resilience is a complex task, and the need to do so has produced researches that aim to classify urban resilience into sub-indicators or resilient components. Researchers perhaps provides the foundation for classifying urban resilience into four main indicators: namely, infrastructural resilience, institutional resilience, economic resilience and social resilience. According to researchers, infrastructural resilience is associated with lessening the vulnerability of critical urban facilities to hazards, whereas institutional resilience on the other hand relies on the protection and strengthening of governmental and non-governmental systems that manage an urban area. Economic resilience deals with improving a city's economic diversity such as employment diversity, whereas social resilience relies on enhancing a community's social capital and the community's sense of place.

Researchers promote a fifth component: "ecological resilience", which assesses the state of ecological buffers, land-use diversity and ecosystem services as crucial indicators for improving the resilience of communities to natural disasters. Researchers, however, make use of the terms "physical" and "environmental" resilience to, respectively, imply the same indicators of 'infrastructural" and "ecological" resilience used in previous researches.

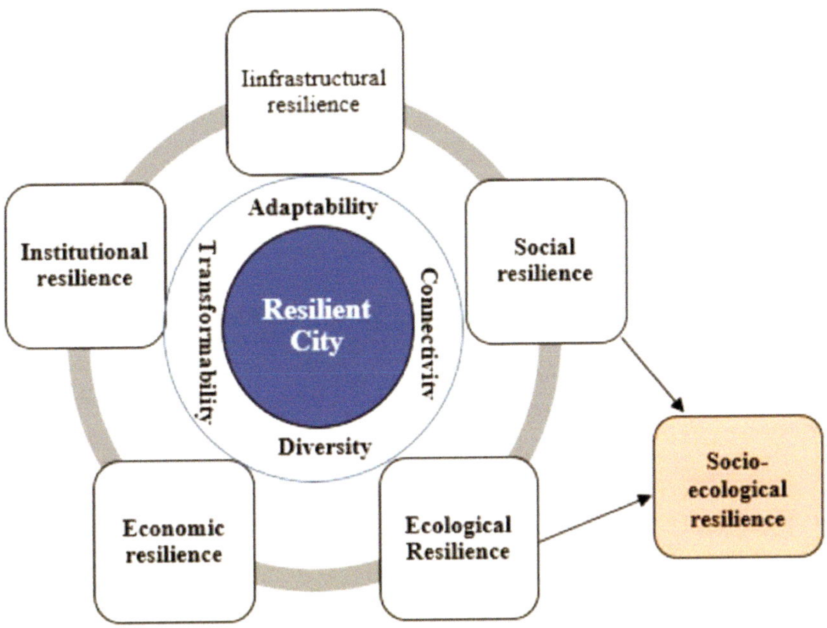

Figure 1. A diagrammatic representation combining the concepts and principles of a resilient city

Researchers on the other hand emphasise the link between social and ecological systems in what is termed as "socio-ecological resilience". The term socio-ecological resilience is used to imply the capacity to adapt or transform in the face of changing social-ecological systems, particularly unexpected change (natural disasters), in ways that continue to support human well-being. The use of the words "adapt" and "transform" in reference to socio-ecological resilience of cities are emphasised as important principles of resilient cities by researchers. Hence, social, infrastructural, economic, institutional and ecological systems in cities are required to possess adaptive and transformative capabilities in order to make them resilient (see Figure 1). “Adaptability” is defined as the capacity of a system to adjust to external and internal change through self-organisation and collective learning, whereas “transformability” is the ability to evolve into new or innovative configurations when current conditions are deemed unsustainable.

In addition to adaptability and transformability principles of urban resilience, researcher emphasises a “diversity” principle for building resilient cities. According to researchers, diversity usually requires redundancy, but having multiple or alternative sources is assumed to provide more security of supply in the face of unexpected disturbances.

Researchers however highlight a fourth principle for building resilient systems: “connectivity” which is defined as the manner and extent to which people, resources and wildlife disperse, migrate, or interact in socio-ecological systems (Figure 1). Connectivity is an important principle because it prevents the fragmentation of urban systems by fostering links and channels where components of resilience can interact.

Assessing resilient cities: The Grosvenor research report

Several studies have attempted to assess the resilience of existing cities or communities based on one or more specified indicators or principles. One such study, the Grosvenor research report conducted by researchers, assessed the resilience capacity of 60 cities (in advanced and developing countries) using verified datasets based on two indicators; “vulnerability” and “adaptive capacity”. Vulnerability was measured by assessing the impact of climate threats, environmental degradation, availability of resources, infrastructure and community cohesion, whereas adaptive capacity was measured based on strengths of disaster management systems, governmental/institutional systems and access to funding.

Assessing the resilience of varying economies in the world was a major strength of the study as it provided opportunities for comparisons between cities with different socio-economic profiles. Three cities in Canada namely, Toronto, Vancouver and Calgary topped the list of most resilient cities based on the set indicators. The weakest 10 cities including Cairo, Dhaka and Mexico City were in developing counties with high urban population growth rates. These cities were revealed to be most vulnerable due to datasets that pointed out to poor infrastructure provision, inequality, environmental degradation and climate vulnerability. The adaptive capacities of these cities were also weak due to issues relating to poor disaster-risk management, unstable political economies and financial constraints.

Toronto is ranked the most resilient city in the report because of the significant strides made over the years to help reduce the city's vulnerability to disasters (flooding) and strengthen its adaptive capacity. Urban green infrastructure is used as a major adaptive strategy for improving socio-ecological resilience in Toronto according to the city's Environment and Energy Division. Proactive adaptive measures used in Toronto include the planting of more trees in the city to help purify and cool the urban temperature; the installation of permeable surfaces (grass and shrubs) rather than asphalt to reduce storm water runoff; and the reliance on drought-resistant plants. The case is similar in Vancouver as green infrastructure feature prominently in the city's "Greenest City 2020 Action Plan", an ambition to make Vancouver the world's most environmentally friendly by 2020. Researcher also affirms the importance of green infrastructure systems such as wetlands, green roofs and buffer zones in reducing the risk of cities to water hazards such as flooding, hurricanes, tsunamis.

While cities in advanced countries such as Toronto and Vancouver are being adapted for disaster risk management and climate change mitigation, cities in developing countries in SSA (e.g. Ghana, Uganda, Ethiopia, Nigeria, etc.) continue to be most vulnerable to the impacts of climate change and natural disasters. Ironically, strategies used by successive governments as resilience interventions in this region (examples are desilting choked drains and the demolishing of structures on waterways) have been reactionary rather than proactive in nature. It is therefore imperative for stakeholders in developing countries to adopt emerging measures that can help strengthen the adaptive capacity and lesson the vulnerability of their cities to climate change. This section contends that, eco-aesthetics presents a fine opportunity for cities in SSA to enhance their urban resilience. Although this section has provided a general overview of the various components and principles of building resilient cities thus far, the focus of this section (which has an urban planning background) will be centred on how eco-aesthetics can be used to strengthen the socio-ecological resilience of cities in SSA.

DISCUSSIONS: POSSIBILITIES OF BUILDING SOCIO-ECOLOGICAL RESILIENT CITES THROUGH ECO-AESTHETICS WITH SCENARIOS FROM GHANA

While the concept of eco-aesthetics seeks to enhance the ecological and aesthetic qualities of green infrastructure systems, it creates opportunities for building "socio-ecological" resilient cities that are capable of adapting to the changing global climate. As mentioned in the foregoing parts of this section, developing countries in SSA like Ghana suffer the most impacts from acute shocks such as floods and epidemics, as well as, chronic stresses such as severe droughts. This section proposes that the principles of eco-aesthetics (ecology, aesthetics, culture and sustainability) will help build socio-ecological resilient cities in Ghana and other developing countries in SSA.

Implementing an "eco-aesthetic" approach in urban planning in Ghana and other developing countries in SSA will imply creating new and enhancing existing urban green infrastructure systems that provide valuable "ecological", "social" and "economic" benefits to cities and communities. These benefits derived from eco-aesthetics can be referred to as "ecosystem services" according to Millennium Ecosystem Assessment (MEA). Ecosystem services are defined as the direct and indirect contributions of natural systems to the environment and human well-being. These ecosystem services are grouped into four main categories in MEA report (2005) as provisioning (material or energy outputs from ecosystems); regulating (promote healthy environmental conditions); cultural (provide ecosystem value to people); and supporting services (provide habitat for biodiversity).

Moreover, researchers argue that urban green infrastructure (parks, gardens and waterways) is an important adaptation strategy for emerging cities. Green open spaces create adaptation opportunities through the "nurturing of conditions for recovery and renewal" after a natural disturbance such as flooding. In addition to strengthening the adaptation capacity of cities, social and economic values of real estate and properties surrounding eco-aesthetic green spaces greatly improve. Aesthetic values provided to people through eco-aesthetics also help ensure the sustainability of such projects over the long term, while cultural and provisioning services providedboost urban social resilience through the creation of "liveable cities" that provide alternative economic returns to urban dwellers.

Finally, global climatic changes and growing environmental problems led the United Nations' member states in 2015 to pledge to achieve a set of SDGs to end poverty, fight inequality, and tackle climate change by 2030. This section asserts that the benefits derived from eco-aesthetics outlined in Table 3 will directly and indirectly help to realise 5 out of the 17 SDGs (Goals 3, 11, 13, 14, and 15) in Ghana and other developing countries in SSA. These five goals (good health and well-being; sustainable cities and communities; protect the planet; life below water; and life on earth) can be associated with building socio-ecological resilient cities, especially in relation to climate change adaptation and disaster-risk management.

The socio-cultural benefits of eco-aesthetics listed in Table 3, for instance, will help countries in SSA achieve Goal 3 (good health and well-being) of the SDGs, whereas the ecological benefits of eco-aesthetics will help promote Goals 11, 13, 14, and 15 at the local and global scales. Building socio-ecological resilient cities in SSA will also help save the limited resources of developing countries that would have been used to tackle problems caused by climate change and natural disasters. Given the numerous benefits of eco-aesthetics outlined in this section thus far, one might expect that cities in Ghana and SSA would adopt these principles in planning directives and future projects. Unfortunately, there exist barriers that may likely limit the implementation of eco-aesthetics in Ghana, and these barriers apparently cut across many developing countries in SSA and are discussed as follows.

Likely barriers to implementing eco-aesthetics in Ghana

This section outlines three main likely barriers that may limit cities in Ghana from fully realising future eco-aesthetic proposals: (1) financial constraints, (2) low level of awareness of eco-aesthetics and (3) restricting land tenure systems. The first barrier, perhaps the most obvious in a developing economy, is the inadequate or limited financial resources to support an eco-aesthetic urban planning strategy. Promoting an eco-aesthetic urban planning strategy would require the planting of more diverse plant species (trees, shrubs, flowers, grass, etc.), enhancing the ecosystem of urban waterways (rivers, lagoons, coastal areas, etc.), and the creation of eco-aesthetic public parks and green open spaces. These initiatives come at some costs of implementation, as well as, the additional costs of maintenance of these green infrastructure systems. To a developing country, this is a major challenge since scarce financial resources are overwhelmed by basic needs of a growing population that cover agricultural and economic needs.

However, existing literature reveals that the success or failure of government policies that aim at sustaining the socio-economic sectors is always dependent on the vulnerability of the environment. When the environment is vulnerable and less resilient, its affects public health, livelihoods and aggravates the poverty cycle. Also, the environmental/ecological benefits of eco-aesthetics do always produce economic advantages. The London i-Tree Eco Report, for example, revealed that the city of London derives GBP 132.7 million (USD 161.5 million) annually in quantifiable ecosystem services such as pollution removal, carbon sequestration and storm water management from urban green infrastructure alone. Another study in California in USA also found out that, for every $1 spent on planting and maintenance of urban trees, $5.82 is generated from the benefits of urban trees. Likewise, researchers assert that, adopting green infrastructure as a flood mitigation strategy can save approximately $6.1 million annually in flood-prone communities in USA. Urban planning authorities therefore should not view financial limitations as absolute barriers to implementing eco-aesthetics, but rather they must view them as opportunities to invest in building socio-ecological resilient cities.

Another possible barrier to implementing eco-aesthetics in Ghana will be the apparent lack of public's awareness of eco-aesthetics and its potential benefits to cities. This lack of awareness of the essential contributions of eco-aesthetics to green infrastructure is a key indicator as to why the concept appears missing in Ghanaian academic discourses as well as urban planning policies at the local and regional levels in Ghana. Interestingly, several studies conducted in the country have revealed that a high percentage of Ghanaians (over 80% of respondents in all cases) claim to be aware of climate change and its effects. This seemingly high level of people's awareness of climate change issues in Ghana is attributable to growing public education via radio stations that transmit in various local languages across the country.

Conversely, careful searches through academic databases for studies on eco-aesthetics or urban green infrastructure as strategies for flood-risk management and climate change adaptation in Ghana appears non-existent thus far. The subject area also appears missing in existing curriculums of architecture and urban planning related courses in Ghana. This section, as part of its goals, hopes to stir up research interests and discourses in the field of eco-aesthetics, particularly, its role in building socio-ecological resilient cities in Ghana and other developing countries in SSA.

The third possible barrier to implementing eco-aesthetics especially at the large-scale level would be land acquisition problems associated with the existing land tenure system in Ghana. The existing system involves customary, statutory (constitutional provisions and judicial decisions), and sometimes religious processes. About 80% of land ownership remains largely in the hands of customary authorities and private individuals in Ghana, with a limited state ownership (public land) of about 20%. Land acquisition process is usually characterised by lengthy, cumbersome and expensive procedures that may create difficulties for potential investors to acquire large parcels of land for large scale projects such as ecological parks. Potential investors may also encounter problems of land litigation over the legal right of acquired lands due to the prevalence of some customary owners reselling land to multiple buyers. Despite numerous attempts by successive governments to reform Ghana's land tenure system (an example: the Lands Commission Act 767 in 2008), land ownership and acquisition still remains a highly contested resource. In the concluding section of this section therefore, we present possible ways to overcoming these barriers to implementing eco-aesthetics in Ghana.

MOBILITY SERVICES

Transportation modes are defined as the means of transportation and their attributes. Individual and collective modes are distinguished. Besides them, so-called transitional modes are spreading. The attributes of transitional modes are to be defined between that of individual car use and traditional public transportation. Typically, either the time frame of a vehicle or the available seats of a vehicle are shared. The sharing facilitates the efficient allocation of the available sources. Either demand-responsive or demand-driven (on-demand) mobility services are provided. In demand-responsive services, flexible schedules are applied, and the capacity reflects the demands. Demand-driven services are operated only if any demand arises. The routes and schedules are not predetermined. At these modes, the ordering (demand announcement) in advance is usually mandatory. The real-time demand-capacity coordination and management of the services are based on an advanced infocommunication system. Several transitional modes are available only via a smartphone application. Mobility services can be defined as joint activities that provide the transportation capacities, technological processes, and related (information) services (e.g., seat reservation). Description of typical urban transportation modes and services is summarized in Table 1.

Table 1. Description of transportation modes

Transportation modes	Description
Public transportation	Supply-based services which are accessible for anyone at a specified location (stop), at a specified time (according to timetable), on a specified route (line), according to the specified terms and conditions, for a specified fee
Demand-responsive transportation	Flexible mobility service in space and/or in time which can be used only after request, often with the same tariff as the traditional public transportation has. It is operated typically in a sparsely populated area or in time periods when travel demand is low
Vehicle-sharing (bike-/car-sharing)	Shared use (in time) of a publicly or privately owned vehicle. The vehicles are available for anyone for a (time-based) fee. It is typically used for a short-range, urban ride, for a short term
Ride-sharing	Capacity (seat) sharing of a privately owned vehicle to share the travel cost. Typically, a slightly regulated, nonprofit transitional mode for a long-range travel. The driver offers the free seats of the vehicle
Taxi	Strongly regulated profit-oriented transportation mode typically for a short-range, urban ride. It provides a door-to-door, high-quality ride for a high fee
Ride-sourcing	Application-based ride management. Travel demand and capacity are coordinated automatically. The mode is preferred for a short-range, urban ride, like the taxi service. It provides door-to-door service with slight regulation, applying dynamic tariff system
Chauffeur service	Traveling in our own car as a passenger with a "hired" driver. Typically, it is used for the sake of convenience, usually at night, when the traveler is not capable of driving or wishes to do other activities during the travel
Individual car/motorcycle/bicycle	Using a privately owned vehicle for an individual trip; neither the vehicle nor the seat is shared with unknown travel mate

Walking is incorporated into every trip. Access to the departure point and egress from the destination point, as well as transfer between the modes, require walking. It is competitive for a short range. The so-called soft mobility forms cover micromobility, namely small-sized (one or two seats) human or electric-driven vehicles (e.g., bicycle, pedelec, Segway, scooter). They are used for the last miles. These vehicles are operated in shared mobility services as well.

Transportation modes and mobility services can be categorized according to several attributes. The knowledge of these attributes is necessary to sufficiently serve the travel demand and remain competitive. The most important categorization aspects of transportation modes are as follows (Fig. 2):

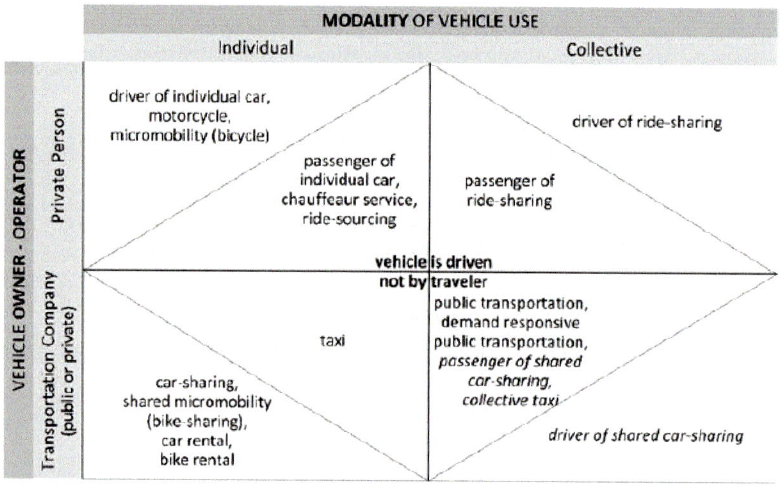

Fig. 2. Classification of current passenger transportation modes

- Modality of vehicle use (individual or collective)
- Vehicle proprietor (service company or private person)
- Driver (traveler or professional driver)

The selection of a mobility form for a specific trip is influenced by several aspects at the same time. Transportation modes can be compared, among others, according to travel distance, the regularity of traveling, the number of passengers, and flexibility. Detailed flexibility analyses focus on the following attributes:

- Spatiality—e.g., departure and destination point, coverage, availability, travel distance, and the fixity of the routes
- Temporality—e.g., accessibility (operation time), timetable, waiting time, service time, time units of the use, and the minimum time of use
- User target group
- Attributes of ordering
- The degree of spread
- Type of the applied infocommunication system and services
- Attributes of the fee collection system—e.g., method of fee calculation and

payment

- Need for additional tasks—e.g., maintenance and repair
- Type of operation—e.g., publicly or privately owned vehicles.

The attributes influence modal share and travel chains. Modal share is a set of indicators which show the percentages of travelers using particular transportation modes in terms of either the number or the length of the trips. Usually walking, cycling, car use, and public transportation are considered. Although the use of transitional modes is constantly increasing, their modal share is still exiguous.

Travel chain is defined as the combined use of different modes and services during a trip. The aims of forming travel chains are:

- To extend the supply of passenger transportation
- To combine the partial advantages of transportation modes

During the formation of an integrated travel chain, the combination of transportation modes is to be planned to provide similar mobility opportunity and service level than that of individual transportation. If a travel chain including a public transportation service provides favorable mobility opportunity according to flexibility and service quality, the share of public transportation can be enhanced. The "weakest link" determines the "strength" of a travel chain. The transfer is the critical part of a chain. The transfers between means and modes are performed at stops and in stations as well as at intermodal junctions. The quality of the mobility service is influenced by, besides the quality attributes of each mode, the design of the intermodal junctions (e.g., walking distances) and the quality of the available supplementary services (e.g., information provision). The common targets during the combination of transportation modes are to minimize the time spent in the system and the fee to be paid while maximizing the perceived service quality. Travelers' movements should be managed in the network considering these targets. The coordination of passenger flows is represented in Fig. 3.

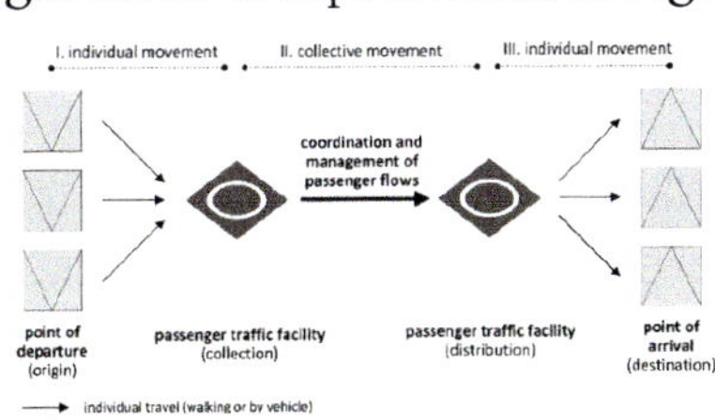

Fig. 3. Model of travel chain

Shared or transitional modes provide services complementing traditional public transportation. So, new alternatives to individual vehicle use are established. The relevance of that is high as the individual cars are used for only 0.75–1 hour per day and only 1.5–1.8 travelers sit in a car on average during a trip in the European Union.

However, the utilization of a vehicle can be enhanced by car-sharing and ride-sharing. The daily utilization of a car can be enhanced to 6–8 hours by car-sharing and the seat capacity utilization to 3.5–4 travelers by ride-sharing.

ALTERATION OF MOBILITY SERVICES

In the early days, walking and animal-powered (mainly horses) transportation meant the only way of transportation. The individual serve of increasing travel demand was not efficient over a specific volume. Thus, public transportation services were implemented. In an urban environment, the horse-drawn omnibus, later the electric (e.g., tram, subway) and internal combustion engine-powered vehicles (e.g., bus), meant the base of transportation.

The use of electric drivetrain remained significant in guided public land transportation, namely, electric locomotive, tram, subway (metro), and trolleybuses. Public transportation may significantly contribute to the decrease of fossil fuel dependency. Furthermore, nowadays electric vehicles are getting more and more attention in the field of road transportation.

Because of economic growth, changing lifestyle, and technological development, mobility demand is increasing. Private car ownership and private car use have grown by leaps and grounds. The increasing number of road vehicles made imperative the management of traffic flow. High-quality public transportation and the spread of transitional modes can reverse the tendency. Where the road network, mainly in the European cities with historical urban structure, cannot serve the huge volume of private car use or where the economic growth was not sufficient, the nonmotorized and public transportation become widespread.

At early stages, mobility service providers managed the processes independently without cooperation. In the second half of the twentieth century, many transportation centers have been established to coordinate the public transportation services in urban areas (e.g., common tariff and information system, fee collection). The integration is an expectation from travelers as well.

Innovation in the transportation system is motivated by altering traveler expectations and technical developments. The adoption of novel technologies is a solution to several challenges albeit the shift in technology causes difficulties during the implementation phase and the fears around new technologies are significant. As a result of technology development, the characteristic of vehicles and traffic management alter.

Furthermore, the attributes of mobility demand and quality expectations are also altering. The travelers increasingly wish short preparation and travel times, pleasant circumstances during travel, as well as reliable and personalized information services. Intelligent and smart systems are available. Smart systems "know" more than intelligent systems because they are able to learn. The learning ability is programmed into them. With learning ability smart systems can respond not only predetermined situations.

Vehicle automation generates alterations in planning, operational management, and passenger handling functions. The current transportation modes are usually operated under human control, while the back-end functions are more and more automated. In general, automated and autonomous systems are distinguished:

- The automated system is controlled by computers following predetermined, step-by-step rules which were programmed into them, respectively. The algorithms cover potential situations and their consequences.
- The autonomous system is controlled by computers which make individual decisions using cognitive capabilities and learning capacities to manage the situations that have not been known before.

Automated and autonomous vehicles can be considered as moving computers. Several pieces of hardware and software are required for the operation. The operation of these components is simpler if the propulsion of the vehicle is electric. Moreover, the control functions needed for driver-less operation can be adopted easier in an electric vehicle. The main development areas in automotive technology are vehicle control, communication system, validation, and verification of components [8]. In the automotive industry, SAE levels are commonly used for describing only the vehicle control.

Several types of public transportation means are already automated, especially in track-based services (e.g., subways). Personal rapid transit (PRT) and group rapid transit (GRT) are relatively new mobility services operated mostly by automated vehicles. The comfortable, exclusive, either feeder or point-to-point services are available in small-scale areas in order not to give up the advantages of individual travel.

The entire mobility system is expected to be reshaped due to the widespread use of fully autonomous vehicles. According to the expectations, a high proportion of road vehicles becomes automatized. Currently, the developments of AVs focus on most types of road vehicles such as car, small bus (so-called pod), as well as even air vehicles (drone).

Current transitional modes are merging into a new mode—so-called shared AV. This mode is infocommunication-based, highly personalized, shared, and available via mobile application and provides mostly on-demand or demand-responsive service. Small capacity vehicles, so-called pods, are applied. The seat or trip reservation is mandatory, which supports capacity planning, and so results in efficient transportation. The reservation, especially in urban transportation, requires additional action from the traveler. Since the large one-directional travel demands cannot be served efficiently by any other modes, the high-capacity, arterial public transportation lines remain important in the future too. Different scenarios are made for the type of this mode:

- Door-to-door, shared service
- Feeder service to high-capacity public transportation
- Combination of the previous types

Personalized, flexible, door-to-door service can be provided mostly in rarely built-up areas because of the road capacity limitations. Consequently, in densely populated and highly urbanized areas, feeder service should be provided. The demands can be influenced by the application of dynamic tariff structures. In this way, the traveler is motivated to use the feeder and shared services.

Based on the scenarios, we defined the types of shared AVs:

- Taxi provides individual door-to-door service between any departure and arrival points without sharing the seat capacity.
- Shared taxi provides door-to-door service between any departure and arrival points with sharing of the seat capacity.
- Feeder pod provides feeder service from any departure points in a zone to the stop of an arterial, high-capacity line; transfers are guaranteed by semi-fixed timetable. The operation is symmetric in the opposite direction (from the stop of an arterial line to any arrival points).
- Fixed route pod provides mostly feeder service on fix route. The departure and arrival points are fix stops. It is operated according to fix timetable, but additional departures may be inserted according to current demands.

This mode can serve a significant rate of demand derived from individual car use. The individual, private AVs are used only for the most flexible travel purposes. The proportion of soft mobility forms, i.e., walking, and micromobility remain significant in the cases of both individual and shared use. However, the circumstances of soft mode use should be improved, especially in the downtowns.

The modality of the vehicles tends to collective modes due to shared AVs. Furthermore, most of the vehicles are to be owned by a company. The mobility services are managed by an integrated mobility service provider, while vehicles are owned and maintained by either a transportation company or a private person. Especially the smallest capacity vehicles used for taxi and shared taxi service may be owned by a private person. These vehicles can be publicly used when they are not used by the owner (i.e., ride-sourcing service with AVs).

The future categories of passenger transportation are:

- Individual transportation
    - Nonmotorized: walking and cycling
    - Motorized: individual AV, motorcycle, and micromobility
- Public transportation (mobility services)
    - Small capacity
    - Nonmotorized: bike-sharing
    - Motorized: shared AV, other shared micromobility
    - High capacity (mass transit) based on AVs (e.g., bus, tram) or highly automated vehicle (e.g., subway)

PLANNING PRINCIPLES

The new mobility services require new approaches in planning. Especially, knowledge about mobility services based on AVs is still lacking as vehicle developments are still at an early stage. However, the various and combined service types require novel comprehensive planning and operational methods. Traditional methods should be altered as the consequence of the following factors:

- Complex system architecture (e.g., the vehicle becomes an independent system component, and the number of operational functions is increasing)
- New and unknown technology
- Increasing dynamism of data management
- Altering user preferences and expectations toward a higher quality of service

The traveler needs to learn completely new tasks, and the existing ones should be managed in a different way (e.g., ordering, open/close the vehicle, payment). Because of AVs, the role of human staff can be reduced, and drivers' work regulations are not to be considered anymore. But the lack of staff attendance implies several new challenges both in management (e.g., electric energy charging) and passenger handling (e.g., safety, information). Moreover, the vanishing driver as a job will cause a societal challenge.

Most of the functions can be automated either partially or entirely. The most relevant challenges are revealed in the following functions:

- Real-time demand-capacity coordination
- Planning vehicle runs with and without passengers; furthermore, shared runs are to be introduced
- Customization of mobile application and supplementary services

- Automatic real-time, personalized, and location-based push information provision
- Automatic vehicle charging

In advance mobility services, both data collection and planning functions are supported by software. The planning and operation of the novel, transitional services and shared AVs require a high amount of real-time data. But, as the technology is new, operational data are unavailable. The collection and consideration of user expectations are inevitable. The acceptance and easy adoption of a new service or technology, like AVs, can be significantly enhanced by a highly personalized mobility service considering the user expectations.

Not every traveler group can be served by the new mobility services. For instance, the ride-sourcing services are used by mostly youngers with high education background and for short distance. The acceptability can be measured before the use (as an expectation), whereas the acceptance itself is to be measured after the use (as a revealed preference). Numerous publications deal with the measurement of expectations toward services based on AVs mostly by stated preference questionnaires or in some cases by revealed preference questionnaires. Measuring the acceptance is rather difficult as bare experience is available. Consequently, the expectations can only be measured according to the stated preferences. But the acceptability is predominantly influenced by the perceived usefulness, expected effort, ease of use, and social influence.

In our previous studies, we performed a questionnaire survey to reveal user expectations. The main findings were:

- The preferred service type is influenced by motivation and current mode use. The less-flexible (pod-like) types are preferred for less-flexible motivation (e.g., work/school), whereas the flexible (taxi-like) types are more likely preferred for the ad hoc travel motivation (e.g., leisure activity). Current car users prefer flexible types, while public transportation users accept less-flexible service types.
- Higher spatial coverage is expected from the shared AV mode. Namely, the willingness to walk is less in general than in the case of a traditional bus service. But it depends also on the current mode choice preference, the size of the city, and age. Young generations living in big cities and using public transportation have a higher willingness to walk.
- The more flexible the service is, the higher the acceptable fee level is. The younger generation's willingness to pay is higher.
- Mobile application functions that help the use of the mobility service are the most important (ticketing, entitlement checking, fee calculation, display travel-related information).

- Services providing travel-related information, along with the services supporting the use of individual smart devices (free Wi-Fi, chargers), are the most important onboard services.
- Activities which are limited or impossible to do during driving become more popular in the future (looking around, surfing on the Internet/chatting, working).

IMPACTS OF ALTERATION

Reshaped urban mobility implies several impacts. Due to shared and demand-driven services, it is expected that the individual car use will decrease. However, the change in the travelers' mindset is needed. In an urban area, individual car use and car ownership are to be eliminated. Necessarily, efficient public transportation and introduction of shared and transitional transportation modes are required. In developed countries, the intention to own a car is decreasing among the younger generation.

Significant benefits can be gained by dynamic assignment of shared vehicles for different service types. The benefits are relevant for passengers (e.g., reduction of travel time and fee) and operators (enhanced utilization rate). However, the utilization of shared services is not efficient without adequate management. For instance, the capacity utilization of a taxi service is not efficient without seat-sharing. The mileage of taxis can be reduced by 40% by seat-sharing [28].

Moreover, if shared, demand-driven services are too convenient, the share of high-capacity public transportation decreases, which may cause further problems (e.g., increasing traffic jams). For instance, at least half of the ride-sourcing trips replace traditional public transportation trips.

The application of AVs has impacts on society and the environment too. The length of trips is expected to grow as travel time is spent in a more efficient and pleasant manner. In this way, daily commuting distance can be even longer. Furthermore, travel time can be reduced as a result of better drivability of the vehicles and advanced traffic control. Vehicle design and passenger compartment also alter. The individual total utility of travel increases, because onboard activities may replace activities performed at home or the workplace. All in all, the number of travels is expected to increase.

However, individual car use decreases by the introduction of a flexible shared AV service because similar service quality can be provided as by an individual car. Current car users' willingness to shift is the highest, as bikers' and pedestrians' willingness to shift are the lowest according to our previous study. The number of cars decreases as less vehicle is enough to meet the needs. Furthermore, the capacity utilization and useful mileage of the vehicles increase because of the shared use.

However, the number and length of empty runs may increase due to dissimilar pickup and drop-off points of passengers.

The vehicles communicate with each other, with the infrastructure and with other road users. Consequently:

- The number of road accident decreases, and traffic safety increases.
- Traffic parameters alter, for instance, smaller headway and higher speed.
- Traffic control alters, for instance, less traffic sign is enough, but a complete replacement of them is not expected as they are needed by soft mobility users.

The public places in cities are utilized in a different way which causes a paradigm shift in urban design as well. The number and extension of road infrastructure elements are also expected to be altered. For instance, less parking lots are enough. Accordingly, the available space for pedestrians and micromobility users or other, non-transportation-related functions (e.g., green areas) can be increased. The time base of road infrastructure elements is to be shared between the functions:

- Shared parking lots:
  - At night or daytime, residents or shared AVs park and recharge.
  - In daytime shared AVs use them as virtual stop.
  - In dedicated time intervals, freight AVs serving neighborhood shops, restaurants, etc. park and recharge during loading.
- Shared traffic lanes: in the peak hours for moving traffic, in the off-peak hours for parking

Though the AVs may serve any point, they cannot stop "anywhere." Therefore, allocation of virtual stops is needed. Virtual stops are points where passengers can board and alight an AV in a safe way and the traffic is not hindered. Virtual stops can be established without any special infrastructure.

The energy consumption becomes more efficient as a result of energy-efficient vehicle control. Thus, the environmental impact of transportation is also reduced. Pollution decreases further if electric and renewable energy sources are applied.

The traveler groups alter as well. Current car drivers become passengers. Demand-driven mobility services provide better spatial coverage and time availability as they usually serve door-to-door rides on demand. Accordingly, some pedestrians become also passengers as the use of a demand-driven service can be faster and more comfortable than walking.

The small AVs may be also considered as an accessory of a building. While the elevators support the vertical movements, the small AVs connect the locations with horizontal movements. The travelers can enter the building with them (e.g., during shopping). Thus, the concept of smart vehicle and smart mobility is related to the concept of smart home or building, and from a wider perspective, these all are part of the smart city concept. However, all in all, the share of pedestrians and micromobility users is expected to increase as soft mobility modes are promoted, especially in urban areas (e.g., pedestrian zone). Accordingly, both soft mobility modes and the shared use of AVs as a feeder service are needed in order to avoid the significant increment of motorized road traffic.

CONCLUSIONS

In the effort to better understand human decision-making processes, both individual and collective, philosophical and applied aesthetics can provide insight into how aesthetic values influence these processes. It is important to study how personal, individually executed aesthetic choices manifest in the everyday practices of contemporary societies when determining the significance of aesthetic values for sustainability transformations. Aesthetic preferences—or more broadly, taste—are never purely subjective, but are formed in a complex network of personally developed and even biologically determined tendencies to be attracted by something that is intertwined with what is socially valued, acceptable, or avoided. To some extent, the most commonly shared aesthetic preferences seem to reflect the general value ethos of their time. With this in mind, it is not an exaggeration to state that we are currently witnessing the formation of a new aesthetic ethos negotiated through the terms of sustainability.

It seems clear that people will still also continue to enjoy (aesthetically or otherwise) things that are not good for them, others, or the planet; such destructive human behaviour is not satisfactorily encountered by prevailing contemporary scientific paradigms, which tend to emphasize the rational side of human activity. In this sense, the humanities are of crucial importance to the development of sustainability studies, since it will be through history, language, narratives, representations, and art that we will have at least some possibility of understanding the darker tendencies of the human societies and the overall processes of how human values develop.

CHAPTER

SCALE ESSENTIALS 7

Buildings consume the largest amount of energy in cities and simulating urban energy dynamics provides the most cost-effective references for urban building planning and energy policy-making. However, cities have a tremendous number of buildings and complicated physical/environmental conditions, current simulation models require formidable computation resources and time. This section proposes a rapid simulation approach that decomposes city model into spatially correlated building blocks for distributed simulation. The proposed distributed adjacency blocks (DABs) algorithm utilizes 2D footprint to construct 3D building groups and solar azimuth angles, altitude angles, and shading plane to simplify simulation targets. With the proposed method, the energy dynamics of the whole city can be simulated in parallel with multiple threads through abstracting inter-building boundary conditions. To validate the proposed method, this study conducted two validation experiments with different building numbers, window-to-wall ratio, and climate conditions. The simulation results suggested that the proposed algorithm can dramatically improve the simulation efficiency and generate less than 5% of percentage difference compared with the conventional whole city simulation approaches.

The cross-examination of the relationship between a city scale and architectural scale of buildings that form part of an urban team is one of the fundamental issues underlying the conduct of original, creative and well-thought-out design performance. The ability to operate scale has been a domain attributed to architecture and urban planning designers for centuries. The scale is not a uniquely defined term, but commonly used, usually intuitively and based on subjective judgment. In general, this is not a problem unless we consider the scale as a criterion for verifying the effects of building creativity. The absence of a clear definition of the scale in such cases proves to be a definite flaw. Therefore, the reference of the term of scale to the scale of an observer seems to be the most appropriate. To simplify, an adjustment of space to human needs refers essentially to two problem areas.

The first, considerations over modelling buildings and their environment was developed in relation to aesthetic human sensitivity. The second, more contemporary, extends the scale issues to human mental needs. Tracking of the development of the issues above provides some perspective on the current views on the creation of the building environment.

ARCHITECTURE AS GEOMETRICAL AESTHETICS OF PROPORTIONS

In architecture, the method of shaping buildings based on proportions has had a long tradition. The question of proportions, as a way of organizing spatial reality, has already been known in art since ancient times as the relation of particular parts to the whole. Irrespective from its definition, a proportion is the mean by which a construction work may be divided to produce the effect of rhythm, harmony, balance, unity, and contrast. In this case, it is also possible to reverse the action combining the composition elements into a coherent whole.

The proportionate meant understandable, simple and good, and therefore considered ideal and beautiful. Forms occurring in nature were generally considered worthy of imitation. The figurative images of palm, papyrus or lotus shapes (Fig. 1), as well as human Fig.s and zoomorphic accents (Fig. 2), are inextricably associated with the ancient Middle East architecture. The aesthetic solutions were sought in this way constructing the monumental buildings in Mesopotamia, Persia and Egypt.

Fig. 1. Fragment of the Egyptian column with a stylized capital in the form of a lotus leaf, The Temple of Amon in Luxor, Egypt, XIII B.C.

Figure 2. Capital of the column in the zoomorphic form (protoma) from the apadama of the palace in Persepolis, Persia, V B.C.

Figure 3. Doric order, the Parthenon in Athens, V B.C

The continuation of the Middle East heritage can be fully observed in the ancient Greek architecture, the first of the ancient traditions to have been comprehensively identified. Significantly, the aesthetic sensitivity of the Greeks went hand in hand with the ability to describe it by means of the principles of Euclidean geometry. The ability of abstract modelling of forms, originating from the nature and wooden construction tradition, has allowed them to be transposed into a more durable stone material. The ability to operate a proportion resulted in the development of architecture and art in the classical period (V- IV, B.C.), both in the field of construction (Fig. 3) and in theory ("golden ratio", aesthetic canons and orders of architecture).

The Hellenic method of "beauty codification", recognized as a creative tool, has been an inspiration for the construction activities development for successive generations to this day. The assessment of the Greek achievements limited only to the admiration of the mathematically aestheticized maturity of the design problems solving is not complete in this case. At the same time, non-aesthetic aspects related to viewer mentality appeared too. The viewer perception resulting from natural inclination (human eye structure) to emphasize straight lines, now recognized as Zöllner's and Herring's optical illusions was already deliberately corrected by the subtle deviations of some structure elements to "straighten" them horizontally and vertically. Moreover, the anthropomorphic features (Doric - male, Ionic and Corinthian - female and girls) were assigned to the architecture orders, and the structure composing was described by the multiplication or division of modules derived from the human Fig. dimension. The Greek anthropocentrism, not only in the material but also spiritual scale, has not been forgotten thanks to the Vitruvius' message (1st century BC), and the concept of the man as the "measure of all things" could get revived a few centuries later.

THE HUMAN SCALE ARCHITECTURE

The issue of architecture proportions and their relations with a human happened to be postponed for many years. Although in this case, it would be more appropriate to say the matter above was simply forgotten. The beginning of the modern era had the issue actively reconsidered at last.

The "Vitruvian man" concept renewal, or rather anew-creation (da Vinci, 1490) was favoured by the Renaissance growing interest in ancient tradition humanism. On the other hand, the preserved legacy of ancient construction performance was the opportunity to recreate its aesthetic canons published in numerous treatises modelled on the work by Vitruvius (Alberti, Vignola and Palladio). However, the resurrection of classical traditions is only a return to art modelled on a flat proportion. It was the analysis of a created work in a perspective illusion that gave the opportunity to model it in so far unknown conditions. The possibility of obtaining a perception similar to a prospective observer proved to be an incredibly useful tool for an architect. The first effects of the linear perspective principles application are known from Brunelleschi's works. Florence's Ospedale degli Innocenti or the Pazzi Chapel are still awe-inspiring with their lightness, harmony and logic of composition (Fig. 4, 5). The buildings are not only the examples of proper handling of proportions, but also above all the skilful adaptation of the whole composition to the aesthetic sensitivity of the recipient.

Figure 4. Ospedale degli Innocenti, Piazza della Santissima Annunziata in Florence, by F. Brunellechi, completed 1424.

Figure 5. The Pazzi Chapel at the Church of Santa Croce in Florence

The Brunelleschi's pioneering concept, however, had its limitations. This was due to the knowledge of the linear perspective only. Hence, his works are dominated by the need to create symmetrical planes and place major artistic accents on the building central axis. This does not change the fact that, for the first time, the creator went successfully beyond the rigid corset of a flat proportion and took into account the conscious presence and therefore, the scale of the future observer. Certainly, these were only humble beginnings. The majority of architecture structures were still perceived as separate and finite entities, and the creative process was dominated by the use of more or less complex systems of proportions.

ARCHITECTURE IN THE URBAN SCALE: TOWN PLANNING

The wider interest in mutual relationships between architecture structures appeared in the Baroque period. The buildings began to be watched, not only from one but from many directions. It mainly concerned view axes which future works would be viewed from in reality. The observation method of architecture as a sequence of scenes and images, first designed and then viewed by the audience was introduced into designing.

Buildings were not created as individual structures, but rather as view perspectives as stage systems of which the buildings are parts. However, it was not about the construction of unified urban complexes without distinct individual features. On the contrary, the analysis of the selected fragment of the urban "scene" was used to deliberate introduction of a hierarchization of buildings among which the selected objects either dominated or were merely the background for others. This way of space organization satisfied the needs of the most active investors of the epoch: The Church and the state. Hence, the creative effort of architects focused mainly on the design of monumental and effective works aimed at emphasizing the prestige and adding up to the splendor of papacy and secular authority. Paradoxically, the impulse for the development of the Baroque vision of the city were not some lofty ideas or aesthetic concepts created by architects but a realistic assessment of spatial and functional reality and the need to undertake regeneration actions, in this case for the falling Roman metropolis. The decision to put the chaotic buildings of the medieval old town in order initiated by Pope Sixtus V (1585- 90) was definitely utilitarian (Fig. 6). However, it must be emphasized the visionary myth often attributed to Sixtus V. of deliberate introducing novelty composition solutions into urban planning is definitely an oversimplified message.

The foundation of the papal actions was not a creative statement, but a necessity to solve the current problems faced by the city and at the same time the headquarters of the church authorities. Apart from the prestige and sanitary issues1 , the metropolitan administrator expressed clearly the utilitarian priorities. Numerous group of pilgrims coming into Rome were to follow clear circulation routes towards the main pilgrim basilicas (San Trinita dei Monti, Santa Mari a Maggiore, San Giovanni in Lateran and di Santa Croce in Gerusalemme (the Holy Cross in Jerusalem)), thus avoiding getting lost in the alleys of the capital as it used to happen. The vertical elements set in front of temple facades were to add up to facilitating the spatial orientation; the Egyptian obelisks were commonly imported to the city even in the glory days of Imperium Romanum, the Roman Empire. Moreover, due to the large number of pilgrims gathering before the liturgy, there was a need for an adequate space free from buildings at the main church entrances. This much resulted in a new look at the problem of shaping urban space.

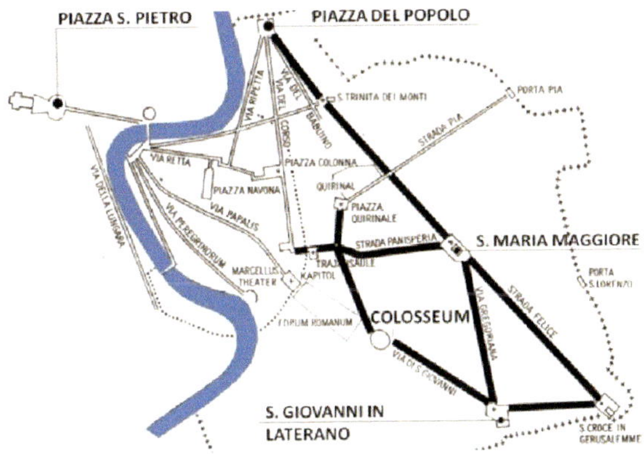

Fig. 6. The new plan of Rome - the course of the new street regulation introduced by Sixtus V (1585-90) is marked in black.

The necessity of introducing a new street regulation in Rome provided an opportunity for a clear display of the churches mentioned above. It also facilitated the identification of urban space from the passer by perspective and, most importantly, contributed to the development and future creation of the public space: the essence of the city. New squares of varied geometry were created (circle, oval and rectangle) with branching of streets most frequently in radial pattern (Fig. 7).

Fig. 7. Piazza del Popolo in Rome - an example of urban space order initiated by Sixtus V in 1589, the present oval shape of the square dates back to 1822

The bodies of churches were exhibited in the building complexes (Fig. 8). They were superior to subordinated axial symmetry and urban interiors symmetry strengthened by placing various elements of development near the temples. The decoration of the squares was often supplemented by the effective gardening designs, fountains (Fig. 9), pools, wells, monuments and stair.

Fig. 8. Piazza San Pietro in Rome, view towards St. Peter's Basilica. - dominant landmark among the accompanying buildings

Fig. 9. The Trevi Fountain in Rome - typical Baroque decor of public space, the fountain element located on the square is the stage design "pretending" façade of two independent buildings

During the papal Baroque era, Roman structures were to arouse awe, admiration and be a testimony of the investor's prestige. Thus, they were supposed to be monumental and so they are. The intentional monumentalism, however, can be attributed to individual buildings and not the public space involved. The buildings at The Piazza del Popolo, Piazza Navona, Trinita dei Monti (Spanish Steps) or Piazza San Pietro are undoubtedly impressive and majestic.

However, there is no manifestation of spatial tyranny against the user. There is the correlation of horizontal dimensioning of squares (length and width) with its walls (height of buildings), pattern and floor planning, selection of divisions and details on elevations of buildings, positioning of vertical accents and the analysis of foreshortening and directions from which buildings would be viewed. All these are simply selected elements of an architect's skills, which influenced the design decisions that made the work created subjectively acceptable to its user.

It does not matter that the church looked like a giant to the viewer standing in front of it, and the column base at the front elevation, of e.g.: St. Peter's Basilica equalled half his/her height. The dimensions of the buildings are secondary here. What seems to be primary here is the manner in which we tune with them. If the general outline of the structure can be seen from a long distance, and coming closer to it, we can see smaller divisions on the elevation until we recognize the architectural detail placed on it, we are able to accept its exceptionally large cubic size limited only by the technical capacity of the epoch.

There is a lot of matters for which architecture, especially Baroque, can be blamed. Theatricality and pompousness of aesthetic solutions, breaking with the classic canons of the Renaissance beautifulness and the multitude of forms present at the same time and space (horror vacui) are just few flaws of the Baroque works.

Without any doubt, you can agree with them, but not in the case of an attempt made to order urban space. Baroque builders can certainly be called urban planners who managed to balance the harmony between the architecture of a building and the fabric of the city. The urban structure, inviolable until then, turned out to be the fabric co-creating the architectural structure. Thus, a clear relation between the scale of the city and the scale of the architecture of building emerged.

The scale of the city and the scale of the architecture of building emerged. The Rome-initiated idea of merging and composition arranging of cities found its followers quickly fully flourishing in consecutive centuries (17th-18th centuries). The Baroque practice of urban design was equally applicable to the already existing centres (Paris, Nancy, London and Warsaw) as well as for new developments (St. Petersburg, Karlsruhe and Washington). It is worth noticing the full dimension reached by residential architecture and their surroundings with its flagship palace work in Versailles, modelled on in Europe (e.g. Sanssouci in Potsdam, The Winter Palace in St. Petersburg). Baroque heritage was still present in the urban practice of the second half of the nineteenth century and in the beginning of the twentieth century.

Eclecticism was then the essence of views on architecture and urban planning. Being far from creative explorations, dominated both in theory and in practice. The architect's practice skills and tools in this case were very mechanical in nature, consisting of the combination of architecture styles with the building function. This was particularly evident in public buildings. We cannot help thinking that the discussion how to design was reduced only to the utilitarian character of the building and the issue of developing a proper neostylization for it.

Eclecticism - historicism was the compulsory continuation of the traditional way of shaping cities. The Vienna's Ringstrasse, the new Barcelona or Haussman's reconstruction of Paris are the flagship projects of the era. In European cities, the United States and other locations within the reach of the western civilization, new urban planning and monumental architectural works were strongly present, referring to great historical styles.

European centers found this aesthetics a natural thing. However, eclecticism - historicism implemented in a culturally different environment appeared quite an original concept. The most unusual example seems to be the Amazonian city of Manaus with the Italian neorenaissance opera house.

Apart from unquestionable achievements, especially in reference to multi-dwelling housing construction needs, common evils could not be avoided. The latter include, among other things, an uptown resident overcrowding, bad hygiene and sanitary conditions and a low building standard. Both the artistic stylistic of eclecticism, metaphorically defined as the "costume of the past centuries at the modern ball" and the lack of interest into basic needs of the human scale, the scale of the user of the existing housing complexes contributed to the reform movement at the end of the 19th century.

MODERNISM IN ARCHITECTURE AND URBAN PLANNING

The modernist architects and urban planners' achievements, once welcomed enthusiastically, receive unfavourable opinions nowadays. Main doubts concern the dehumanization of built-up space. The monotony of architectural forms, the building homogeneity and the lack of human factor in the public space are pointed out. The breaking of spatial ties with the historically built up environment, ignoring sensitivity and human perception needs became reality. The initial lofty postulates of the architectural avant-garde referring to the human scale and its relation to the buildings constructed were not reflected in the construction practice.

The revolutionary Bauhaus view of architecture as a field integrally linked to art, psychology, sociology, technology and nature turned out to be just a motto. On the other hand, the continuation of Vitruvian man's concept in Corbusier's new rendition remained just an elegant mental construct limited to the cubic capacity of the designed structure (Fig. 10). Finally, the extremely negative summary of the achievements initiated by the "modern movement" was influenced by the subsequent universal unification of design solutions supported by the construction economy dictate, especially this residential architecture.

For these considerations, it is important that modernism has increased the building size, and none of the contemporary styles so far has not been corrected this cardinal error. The historical principle practice of juxtaposing a structure with a particular spatial situation was abandoned. The cultural sphere along with the traditional relationship of building unity and the scale of the city was left for many years. The "works" designed were to express the artistic aspirations of their creators and to delight the observer with the modern and unknown "uniqueness".

Fig. 10. The Unité d'Habitation, ("Cité Radieuse", "La Maison du Fada", "Mother of All Blocks") - inspiration for numerous housing estates implemented in the second half of the 20th century.

Moreover, the avant-garde assumption of an architect's infallibility in design decisions and the appointment of him or her for an educator of new art recipients and a social life organizer was a mere doctrine. In this case, the modernist dictum was not much different from the simultaneously developing totalitarian architecture trends (the Third Reich and the USSR) and their derivatives (the Eastern Bloc countries socialist realism), and it was even co-creating them.

From an architect's point of view, modernism was not a negative phenomenon, but rather a noble one. The modernist desire to introduce a designer, as the creator of an accomplished structure, makes this occupation attractive and challenging. In non-aesthetic terms, modernist slogans remain up to date and have always been the starting point for reflections on contemporary architecture. It is about the issues of human personal development, human needs and social behaviours, as well as the issues of coexistence of the built environment with the landscape and nature and the use of technical progress in our everyday life ("the second aesthetics of Werkbund").

Build N Scale Buildings From Scratch

Modeling in N-scale allows you to create large detailed scale areas in a small amount of space. For model railroaders or diorama makers this can be extremely useful as it presents the opportunity to build epically, including everything from single buildings to small towns in your model. While there are a large variety of buildings ready for use in N-scale, getting the look just right may require you to forgo the ready-made choices and build your own from scratch. Working from scratch, you can create any building imaginable, from an ancient manor house to an imposing skyscraper. With scratch building you remove all limitations, leaving your scaled areas open for whatever look you're trying to achieve.

Step 1

Locate an image of the building you wish to recreate to scale. You can find images of structures in online photo repositories such as Flickr, or Wikimedia Commons, or architectural magazines.

Step 2

Draw two scaled drawings of the building onto a piece of section. Use a conversion rate of 1 scale inch for every 13 feet, 4 inches of actual building size. Draw each external building surface, using a straightedge where necessary for guidance. Draw only the shapes of each surface for the first drawing. This is what you'll use as a template for creating the basic structure. Fill in as many details as possible into the second drawing. Include labels as to the materials used to create the building with the second drawing for later recreation on scale.

Step 3

Gather the materials you intend to use for the building. You can purchase building supplies made of styrene that mimic common building materials such as brick, concrete and clapboard in scale, from a hobby shop that sales model railroading supplies. Other commonly used scratch building materials are balsa and brazilwood sheets, foam core sheets and cardstock. You can shape these items however you like to create the different elements of your building from walls to rooftop. You can also purchase harder materials to recreate detail pieces such as windows, decorative posts and awnings from the same railroad hobby shops.

Step 4

Use your scale drawings as a template for cutting the body of the building. Transfer the measurements over to the building material sheets with a pencil, or cut the shape from your drawing and place over the sheet to outline it onto the sheet. Cut the shape from the sheet with a hobby knife. Sand the edges of styrene with a fine-grit sandsection to clean any jagged edges from the cutting process.

Step 5

Glue the building pieces together using plastic cement for styrene and white craft glue for other building materials. Hold the pieces together for about a minute before setting them down to dry. Allow the glue to dry completely before continuing.

Step 6

Add details to your building to make it more realistic. Cut pieces of the building materials to create the detail elements, or glue on purchased details. The more detail you add the more realism for the scale model building, but the smaller the building the less realism needed to convey the building image. Use more details for larger models such as large multi-storied homes and office buildings than with small railroad shacks. You can create material elements such as bricks or clapboard by scoring the surface of your material lightly with the hobby knife to emulate the look of the building material. Wait for the added detail elements to dry.

Step 7

Paint the model with acrylic paints to match the look of the real building. Apply a base coat of white paint as a primer to help your top coat adhere to the model. Allow the base coat to dry two hours and then apply the top coat of paint. Use a variety of brush sizes to get the best detail into your model. Wait overnight for your top coat to dry and weather the model to make it appear as though it has been exposed to the elements.

Step 8

Dilute a brown paint with one part paint to five parts water. Dip your brush into the diluted mixture and wipe it nearly dry on a section towel. Lightly go over the surface of the model, drywashing it with the brown diluted paint to give it a weathered look. Apply a second drywash layer for older dustier appearing buildings. Allow the paint to dry for two hours.

Step 9

Apply a layer of matte clearcoat to the model with a paintbrush to protect the paint colors. Allow the clearcoat and the rest of the paint on the model to dry completely for 24 hours before use.

INTEGRATING GEOGRAPHY AND GEOMETRY

Interactions between socio-economic processes and the built environment are relevant to many aspects of urban geographical research. For example, research into urban demographic and economic spatial structures link locational decision-making to buildings through property markets and land ownership. Many significant processes of change in cities such as gentrification and urban renewal involve specific transformations in urban form. Urban sustainability is another research area strongly linked to the built environment, as key aspects of sustainability include the energy efficiency of buildings, and the functional integration of urban activities to reduce travel distances.

Despite these relationships, the direct measurement and analysis of urban form and function is rather limited in geographical research, particularly for city-wide studies. Where urban form is considered in geographic research it is often through aggregate proxy measures such as population and employment density (Talen, 2003). While these measures are a useful starting point, they do not consider any physical properties of built form and leave many relationships unexplored. In this research we argue that the inclusion of urban form and function analysis can provide new insights and empirical grounding for a number of fields, particularly for urban planning and development research, property market and housing analysis, and urban sustainability studies.

There is also a contrast within the fields of architecture, planning and geography in the sense that architects and planners commonly use iconic representations of urban form such as geometric plans and physical models. This approach contrasts with geographical research in two aspects. Firstly the focus is on the physical properties of the built environment rather than the socio-economic interests of urban geography. Secondly the extent of study is generally restricted to buildings and localities, in contrast to the citywide (or larger) scope of geographical studies. There is a growing interest in linking these geographical and geometrical approaches to provide an improved understanding of cities. Over the last decade there has been a continuing development of geographic information (GI) technologies and the emergence of rich fine-scale digital data sources. These new detailed datasets have enhanced spatial and attribute information, and are sufficiently intensive to analyse detailed form and function relationships and also sufficiently extensive to enable patterns to be generalised across entire metropolitan areas. It is increasingly possible to link the socio-economic focus of geographical analysis to the geometric built environment approach that is employed in local urban planning. Researcher has termed this linking process 'geography to geometry', the merging of iconic and symbolic urban models, and it opens up many possibilities for research.

New data models and analysis techniques are required to achieve this goal. This section is intended to further this research agenda: firstly by describing a methodology to combine geometric and socioeconomic datasets for a large city in a single spatial database; and secondly by implementing spatial analysis techniques to provide data and indicators for urban research: principally urban function and residential dwelling type. Fine-scale relationships between urban form and function can then be explored to provide an evidence base for research topics such as urban sustainability, residential property analysis, gentrification, land-use change and neighbourhood definition.

Built Environment Models

In this research we link urban geography to the built environment by integrating socioeconomic spatial data with iconic urban models. As noted earlier, an iconic model is a geometrical representation of a feature or set of features, typically taking the form of a physical scale-model (such as an architect's miniature building model) or a digital model. 3D digital city models have become increasingly widespread and sophisticated with the development and integration of computer-aided design (CAD) software, geographical information systems (GIS), computer graphics, web and aerial sensing technologies. An earlier research project at the Centre for Advanced Spatial Analysis (CASA) developed a 3D digital city model of London called 'Virtual London' and this acts as a foundation for this research.

Potential applications of 3D digital city models range between urban planning, telecommunications, architecture, facilities and utilities management, property analysis, marketing, tourism and entertainment. The development of web and virtual globe technologies has given a massive boost to digital urban models, enabling widespread access and interaction by the public through geobrowsers such as the popular Google Earth.

Whilst the visualisation capabilities of 3D digital city models are clear, their analytical functionality is often underdeveloped (Batty and Hudson-Smith, 2002). Significant advances have been made in increasing the geometrical sophistication of 3D digital city models, but many models remain 'empty shells' without any socio-economic data associated with the buildings or the capability to analyse the role of the built environment in urban processes. We believe that future advances will explore how such models can be populated with socio-economic data and linked to transportation networks thus moving from visualisation to focus on policy applications and analysis. This would essentially mean enhancing digital city models to become Planning Support Systems (PSS), i.e. tools to aid and enhance planning tasks.

We define four levels of integration between iconic urban models and geographical analysis. Starting with the most basic, these techniques are:

- the spatial overlay of thematic data on a city model for visualisation;
- the creation of a city model database with building geometry linked to address/cadastral geography for socio-economic attribution (i.e. a Built Environment Model);
- the combined spatial analysis of socio-economic attributes with urban form geometry;

- the integration of Built Environment Models with urban symbolic mathematical models (for example, land-use transport models).

These methods range from description to analysis, and from static representations to the potential for dynamic modelling. This research focuses on the second and third methods of this typology where the benefits for urban research and PSS are most immediate.

The integration of Built Environment Models with symbolic urban modelling is not explored in this section, though we believe that improved accuracy in the representation of urban form and function would be beneficial, particularly for land-use and urban growth models, and should be investigated in future research.

Fig. 11: Virtual London 3D City Model with Different Data Layers Overlaid on Top. A: Nitrogen Dioxide Layered onto the Street System of London, and B: Querying Buildings Developed from 2001 to 2004

The first method of combining datasets is the overlay of thematic data on a city model. The visualisation of built form and landmarks can help users to navigate and identify familiar urban locations, thus improving data legibility.

Figure 1A illustrates this with an air pollution surface combined with the Virtual London model. Insights can emerge from seeing patterns which may not have been evident without the context of urban form. With this method relationships between urban geometry and geography are visual and lack explicit spatial relationships.

A more advanced approach is to develop a spatial database of the built environment- a Built Environment Model, and store socio-economic attributes associated with buildings. These attributes of buildings can be queried and visualised within a GIS as demonstrated in Fig. 11B where buildings in Central London of a particular age are identified. This type of functionality is relevant to planners querying building stock, and to researchers studying property markets and relating socio-economic trends to the built environment.

To create a Built Environment Model, the geometry of buildings has to be linked to socio-economic attributes through a postal or cadastral geography. Depending on data availability, the development of the data model can be a non-trivial task.

In the United Kingdom (UK) context there has been much recent innovation in spatial address infrastructure, yet a full description of building addresses and sub-building geometry of premises remains incomplete. Data modelling techniques to link topographic and spatial address datasets to try to fill in information gaps are thus pursued in this research.

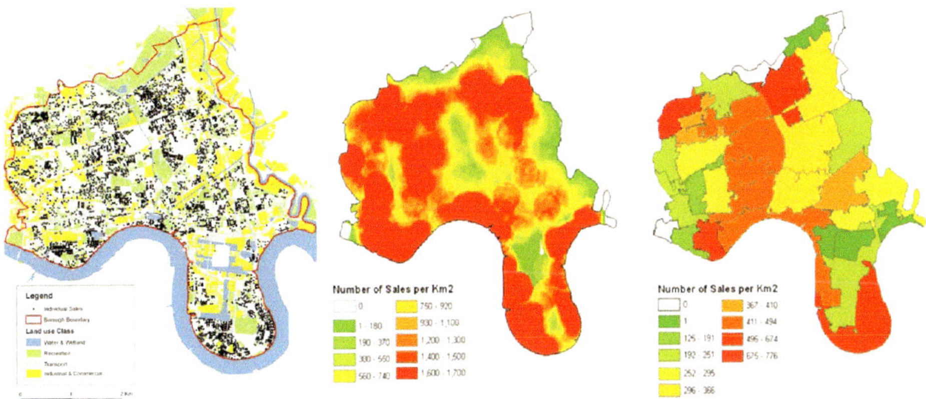

Fig. 12: Masking Variation with Data Aggregation. A: Individual Addresses B: Point Density of Sales, and C: Density of Sales per Postcode Sector.

Once the Built Environment Model has been created then it is possible to perform spatial analysis linking geometric and socio-economic properties- the third stage in our typology of integration. A simple example is to calculate building density measures by dividing a quantity (e.g. the number of residential properties) by building footprint or volume. Any address-based data can be incorporated, from business function data, to geodemographic and neighbourhood characteristics.

Overall with a data model that successfully integrates urban form and socio-economic data, there are a wide range of potential applications relating to planning practice and geographical research. These applications vary in terms of data inputs and analysis methods, and so flexibility is a useful quality of the data model to enable a range of applications.

Scale and Aggregation in Built Environment Models

In geographical contexts the term scale is used to refer to both the level of detail and extent of spatial data and analysis. This dual meaning derives from the common association between data resolution and study area size. This balance is significant both for spatial analysis and visualisation. For spatial analysis large high-detail datasets increase methodological complexity and computation demands; while in visualisation there is a limit to the density of information that is intelligible to the viewer on a page or screen.

The association between large scale and coarse resolution is readily seen in aggregated types of urban measurement, where information on the form and pattern of the built environment is frequently missing (Talen, 2003; Moudon, 2002). Built environment studies ideally require analysis that is both fine-scale, to include premises, buildings and streets, and large extent, to allow the study of city-wide processes. To achieve this, techniques are needed to integrate datasets at fine-scales, make calculations for large urban areas computationally feasible, and to visualise data at multiple scales. The most common quantitative approach geographers use to explore urban data is aggregate zonal analysis. Aggregate analysis is a powerful method of simplifying the complexity of urban data, smoothing over local variation and enabling patterns to be identified. It complements the structure of core datasets such as censuses of population and businesses. Additionally using aggregate data reduces computational demands for spatial analysis operations.

On the other hand, aggregation is a source of uncertainty and error in geographical research. Datasets and zonations can vary widely in terms of geographical scale, aggregation rules and spatial resolution, and this influences research outcomes. All aggregate units are spatially modifiable, i.e. they can be partitioned geographically in many different ways to generate different types of results. This is known as the modifiable areal unit problem (MAUP).

Of concern in built environment research is that socio-economic zonations typically ignore urban form, and overlook variation in physical features. For example, physical urban barriers can be influential in urban processes such as segregation (see Rabin, 1987), yet are often not represented using zone boundaries. To highlight the impacts of data aggregation on the results and interpretation of geographical research we provide a simple example. If we create a point map of housing sales data over several years from a district in London (Tower Hamlet's, for example) we see there is a detailed pattern of clusters spread over the entire borough (Fig. 12A). This detailed pattern can be simplified through aggregation. Fig. 12B we show the point density of individual sales, which captures the general distribution of points whilst losing some of the detail. This contrasts with Fig. 12C, where zonal aggregation (in this instance postcode sectors) has changed the distribution, with the impact of built environment features such as water and parks now masked within the zonal geography.

From this simple example it is clear that the scale and aggregation impacts on the overall results and interpretation.

These issues can be minimised by using disaggregate data. Greater flexibility is possible as fine-scale data can be aggregated into any chosen zonation. Therefore the impact of the MAUP can be tested and quantified. Zone boundaries can be tailored to particular studies, and built environment features can be considered. The effect of the trade-off between level-of-detail and extent can be reduced. This approach relies on the availability of fine-scale geographical data, and increasingly this is becoming available. Privacy is an important consideration for demographic data, and this prevents data such as the census being released at address level. Attempts to disaggregate zonal data are likely to result in ecological fallacy errors, where attributes based on aggregate data are applied to individuals that form the aggregate group. Microsimulation methods are an appropriate tool to mitigate this. In this study we base our analysis on anonymous address-based data and so microsimulation approaches are not pursued, but we believe this could be a fruitful avenue for further research especially in terms of linking iconic to symbolic models in the form of agentbased models, for example. However, the fine-scale large-extent approach is computationally demanding. For a large city such as London, millions of features must be joined to integrate data at address and building scales. The continued advances in computer hardware and Spatial Database Management System (SDBMS) software make this approach feasible.

Suitable data models that fully represent relationships between addresses and building geography are necessary. Once the data model is implemented, then analysis becomes an address matching process (i.e. attribute joins) as opposed to more computationally intensive spatial joins.

SUSTAINABLE URBANIZATION - CHARACTERISTICS OF GREEN URBAN PLANNING

A sustainable city is one designed to address social, environmental and economic impact through urban planning and city management. Many sustainable initiatives are achieved by building eco-friendly alternatives into city infrastructure, such as adopting walk and bike lanes. Regulations and fines can also achieve change: waste removal orders have proven to reduce landfill waste. While finding ways to help the planet is essential to sustainability, reducing costs and creating a vibrant culture for citizens are equally important. Through planned infrastructure, public green spaces, smart waste removal and more, cities can leave behind a net zero footprint for a more sustainable world.

Public transportation

Cities that focus on sustainability must address transportation head-on, as motor vehicles are responsible for 75% of carbon monoxide pollution in the U.S. today. Cost-efficient and accessible public transportation takes cars off the road, reducing harmful emissions generated by daily driving commutes and errands.

Here are some public transit examples sustainable cities have enacted:

- Electric trolleybuses
- Metro systems
- Underground railways
- Maglev trains

Fig. 13. Green Buildings

Another great way to improve public transportation and encourage greener practices is to incentivize alternative travel by reserving certain lanes for buses, electric cars and carpooling. The future of transportation is tackling greener public transit with new technology such as hyperloop and smart roads.

Walkable and bikeable neighborhoods

One of the best ways to reduce carbon emissions throughout a community is to have citizens ditch transit altogether. Walking and biking are great alternatives to encourage, and city leaders can do so by approving infrastructure changes, passing laws to protect cyclists and making smart urban planning decisions.

Take Copenhagen, where bicycles outnumber cars more than five to one, and 42% of residents bike to work. This bike culture was achieved through city management: Cycling superhighways were created across the city by a series of paths and bridges designed specifically for bikers.

Walk and bike-friendly infrastructure includes:

- Pedestrian bridges
- Bike path networks
- Connected urban grids
- Shared micromobility

Vehicle charging stations

An electric vehicle produces zero tailpipe emissions compared to a typical passenger vehicle, which produces 4.6 metric tons annually. For cities hoping to become carbon-neutral, vehicle charging stations for electric or other alternatives must become commonplace.

Cities can incentivize charging station construction by teaming up investors from the private sector with public servants. Urban planners can work with developers to map out charging station locations that suit both drivers and local homeowners. After all, installing charging stations on random street corners could result in property value loss and increased congestion.

Another eco-friendly vehicle that requires newly constructed refill stations is a hydrogen-powered car. Powered by hydrogen and oxygen, these cars can run twice as long as an electric vehicle before needing a refill (rather than a charge).

Solar farms

The switch to clean energy is essential for cities to become sustainable. Solar farms can improve air quality in a city while also providing renewable energy to citizens and municipal facilities. Solar energy also creates well-paying jobs and stimulates economic growth. The U.S. added 110,000 net new clean energy jobs in 2018, outnumbering fossil fuels nearly three to one.

City-funded solar farms allow residents who can't harness solar energy on their own property to gain access to solar power. It also allows municipal buildings to transition to renewable energy. Fayetteville, Arkansas, is working to run 100% of local government buildings on clean energy by 2030.

Green buildings

Thirty percent of the world's greenhouse gas emissions come from buildings. One solution? Green buildings, which reduce CO2 emissions by 32%. They also require less maintenance and have cheaper water and electric bills. These cost benefits, along with additional tax incentives, have allowed green building construction to become widespread without the use of government mandates.

Here are some popular ways green buildings address these areas:

- Smart heating and cooling systems
- Natural building materials
- Enhanced ventilation and insulation
- Solar panels
- Green roofs

In order for a city's infrastructure to be truly green, all municipality buildings should be LEED certified. The U.S. Green Building Council awards LEED certification to buildings that address energy savings, water efficiency, CO_2 emissions reduction and more.

Food production

The World Research Institute estimates that the world will need to produce 50% more food by 2050 while reducing greenhouse gas emissions by two-thirds. This unique challenge requires cities to feed more people via more eco-friendly means.

Fig. 14. Food production

Urban farming is one way cities can cut down on carbon emissions tied to food production. Teaming up with with like-minded organizations can help educate residents on nutrition and encourage them to make healthier, more sustainable choices. Conservation is another area in which governments can invest, but today only 5% of agriculture subsidies go toward these efforts.

Accessible public resources

Sustainable cities also help residents achieve sustainability in their own lives by providing opportunity and access to resources. Public health buildings, economic development centers, technology hubs and more are all examples of resource centers that can stimulate public health and the economy.

These resources also include recreational facilities and programs in which community members can take part. Some of these facilities, such as museums and cultural centers, can rehumanize urban centers and help a city define its identity. Allocating public funds towards cultural amenities and events is a great way to promote diversity and inclusion at the city level and help build bonds within a community.

Water conservation

Resourceful urban water management is also a characteristic of sustainable cities. Over 90% of the world's urban centers are near coastal areas. While this puts these cities at greater risk of natural disasters and climate change threats, it also opens the door for better use of these nearby water sources.

Here are some water conservation practices cities can adopt:

- Rainwater harvesting
- Green infrastructure
- Waterless hardware

Green infrastructure uses the natural water cycle to create clean water sources in place of a water treatment plant. For example, restoring wetlands and planting trees can reduce the need for manmade flood infrastructure such as levees.

Public green spaces

Fig. 15. Public green spaces

Urban green spaces should be a high priority for sustainable cities, as these areas solve a number of challenges at once. Cities can achieve pollution control, public health and thriving biodiversity by adopting green infrastructure, all while increasing property value in the area. Parks, greenways and trails, street trees and protected conservation areas are all examples of public green spaces.

Building community gardens and urban farms can also solve food challenges. In addition to free or affordable gardening spaces for residents, edible landscaping can be incorporated to ensure urban plants serve two purposes: ornamental and functional.

Waste management

Rather than throwing waste away only to end up in a landfill, disposal processes should be circular. While many (if not all) cities have adopted recycling programs, more aggressive urban centers have seen huge gains over the last 20-30 years. In San Francisco, recycling and composting mandates have helped the city achieve a waste diversion of more than 77%.

Energy recovery is a great solution for waste that can't be recycled using traditional methods. This process captures fly ash burned at landfill sites and converts it back into energy.

Using technology to go sectionless is another great way to promote sustainability on a city-wide scale. Many government offices have moved forms and applications online, which is both quicker and less wasteful than mailing or dropping sections off.

CHAPTER 8
SCALES: THE FINE PRINT

The principle of Green Urbanism focuses on zero waste, zero emission, recycling, and natural resources. It aims to bring back a clean environment, water sanitation, and a better quality of life. It is a breath of fresh air when it comes to disciplines approaching urbanism, where cities focus on constructing motorways and prioritizing automobiles as a means of transport, Green Urbanism aims to implement a sustainable means of transportation within cities, that focuses on public transit systems, and the inclusion of pedestrians and cyclists on roads.

When mentioning Green Urbanism, Barcelona's well-renowned superblocks come to mind instantaneously. The cities' structures are an exemplar of green urbanism application. They promote walking and cycling as a means of transport rather than the use of automobiles, thus eliminating intrusive traffic and the pollution associated with it within the block, while the ingenious design promotes the efficiency of public transit systems, due to the density of people living within the superblocks. Verdure and greenery are commonplace within these superblocks, promoting the well-being of their inhabitants and helping mitigate the effects of climate change.

While Barcelona's superblocks offer a dream to strive towards, the mirror image of their quest towards a greener future exists mostly in developing cities, whose population growth outpaced their planning efforts. Within these cities, the urban sprawl has taken hold, causing a rise in crime rate, rampant noise & air pollution, automobile reliance, and difficulty in resource & waste management.

Often times these cities choose to further expand motorways to ease traffic, rather than transitioning away from reliance on vehicles that use non-renewable sources of energy. This, however, is simply an unsustainable approach, that will cost us our continuation on our planet, and cost the cities' inhabitants a livable and comfortable life, introducing a slew of health complications.

The globally distributed health impacts of environmental degradation and widening population inequalities require a fundamental shift in understandings of healthy urbanism - including policies and decisions that shape neighbourhood and building design. The built environment tends to disadvantage or exclude women, children, the elderly, disabled, poor and other groups, starting from design and planning stages through to occupation, and this results in avoidable health impacts. Although these concepts are not new, they are rapidly emerging as built environment research and practice priorities without clear understanding of the interconnected aims of healthy environments that are sustainable, equitable and inclusive.

The Framework was developed through synthesising knowledge from research and practice, and by testing this new conceptualisation in a participatory workshop. Ongoing research is exploring implementation of the Framework in practice. If widely adopted, this Framework may contribute towards achieving the goals of sustainable development through a focus on increasing human health and wellbeing in urban development.

Cities and urban design are intimately linked to sustainability goals. More than half of the world's population now lives in urbanized areas, with these areas accounting for approximately two-thirds of global energy use (International Energy Agency 2012) and expanding into approximately 10 million acres of natural and rural land yearly. With this, the organization and design of urban areas has an ever-increasing role in affecting the sustainability of the planet as a whole.

The link between urbanism and sustainability is well established and is primarily based on the effect of urban form on transport, water quality and recharge rates, habitat, equity concerns and pollution. While sometimes used interchangeably, the term 'sustainable urbanism' encompasses topics of sustainability related to the entire process of city development and management, while sustainable urban design sits somewhere within that.

Sustainable urban design is not necessarily a clearly delineated subset of sustainable urbanism, but instead can be thought of as focus area within it that is concentrated around issues of design while still maintaining strong links to the other realms such as planning, engineering, real estate and policy.

As commented by Alex Krieger, urban design itself is a discipline with an "absence of a simple definition". On the one hand, it "articulates the physical form and programmatic components of urban situations" (Marshall 2009, 40), but the complexity of this task places urban design at the nexus of several other disciplines.

While architects initially claimed the term and discipline, the list of engaged and associated disciplines has expanded due to the complexity of urban development. This list initially included disciplines such as land use and transportation planning, landscape architecture, civil engineering, real estate and public policy. Today, as urban areas tackle sustainability related concerns, the disciplines directly related to urban design have expanded even further to include areas such as ecology, plant and wildlife biology, and environmental engineering to name a few.

Development trend and challenges of sustainable urban design

At present, more than 55% of the world's population lives in cities, and this is expected to rise to 68% by 2050, with nearly 90% of the increase in the urban population occurring in Asia and Africa, especially for India and China. Thanks to the rapid urbanization, the proportion of China's urban population has increased from 17.92% in 1978 to 58.5% in 2017, which is expected to exceed 60% in 2020. The urbanization process has brought about the rapid socio-economic development of cities, yet the rapid urbanization process has also led to some problems related to urban development, such as over-consumption of resources, increasingly prominent pollution, frequent occurrence of 'urban diseases' and extreme weather events, continuous deterioration of living environment quality, etc., which have brought new challenges to sustainable urban design.

Under this background, analysing the basic theories and transformation of urban design, discussing new concepts, technologies and methods of urban design, as well as scientifically recognizing the challenges and opportunities of the development of urban design can provide effective solutions based on urban design for outstanding problems in the rapid process of urban construction and sustainable development.

Urban design and sustainable environment

As one of human intervention practice for urban form evolution, urban design has some common guiding ideology and professional value system in different historical stages, i.e., paradigm.

After experiencing three generations of historical paradigms including traditional urban design, modernist urban design and green urban design, the technical concepts and methods of urban design have acquired some brand-new contents with the increasing development of 'digital earth, smart city, Internet and artificial intelligence', further forming the fourth paradigm of 'digital urban design based on human-computer interaction'.

This new paradigm takes the reconstruction of morphological integrity theory as the goal, the human-computer interaction as the approach, and the transformation of technical methods and tools as the core features. Furthermore, digital urban design can play a key and supporting role in the preparation, implementation and management of urban planning, highlighting the characteristics of urban history, culture and landscape pattern, further understanding and constructing fair and just social norms, and creating a livable and sustainable urban environment.

Characteristics of digital city

Digital city is one of the basic symbols of transformation from the industrialization era to the informationization era, which generally refers to the ability to effectively acquire, classify and store, automatically process and intelligently identify massive data in the category of urban natural, social and economic systems. Taking advanced information means to support the city planning, construction, operation, management and emergency response can effectively improve the level of government management and service, and also promote the sustainable development of city. For digital urban design, the early digital technologies include computer aided design (CAD), computer graphics and virtual reality, satellite images, photographic survey and geographic information system (GIS).

On this basis, the common ways to construct three-dimensional city models can include modelling, simulation and prediction based on CAD design, digital photogrammetry techniques and GIS, etc. At present, the common digital technologies include data acquisition, analysis and visualization. Based on these above-mentioned technologies, traditional data are transitioned to new data (e.g., Internet media), and sampling data are developed into total data with more variety and comprehensiveness, and macro external observation and statistical data are developed into micro individual perception and experience data (e.g., street scenes, internet hotspots, micro blogs, etc.). With the rapid development of digital technology, the innovation and development of urban design technology are constantly breaking through as well, e.g.,

- The overall picture of the world urbanization based on night lighting and density distribution,
- The world map based on the accessibility of traffic time,
- The digital historical hologram map based on historical information,
- The digital maps and city imagery preference information based on mobile phone signalling, Baidu data public preferences and Google Maps, through tourists, residents, professionals or government agencies to participate in the collection of urban imagery, and the formation of urban design needs, etc.

The digital urban design technologies can be summarized into the following three characteristics:

One is the design objects with multiple scales. With the support of digital technology, it is possible for urban design to move towards large scale. In the process of rapid urbanization, the development of various districts in the city can be coordinated and connected, and it has been the general trend to look at space beyond the boundaries of space.

The second is the design method of digital quantification. The digital technology has profoundly changed people's cognition and view on the material form and social structure of the world, which is a brand-new world cognition, knowledge system and method construction in a sense.

The third is the design process of human-computer interaction. Emphasizing the organic integration of values and design creativity with the quantitative integration of multi-source big data has the properties of accommodating development, continuous optimization and error correction. The normal operation of interaction system can be achieved by setting a certain value domain 'fault tolerance'.

CHALLENGES TO URBAN DESIGN AND DEVELOPMENT

With remarkable achievements for the construction and development of digital cities, there are some concerned problems, such as the intensification of digital risks, the similarity of urban forms, the disorder of urban governance, the prevention of sudden disasters in cities, and the large difficulty of multi-scale simulation and design.

DIGITAL INFORMATION AND SECURITY

The digital revolution is the impetus for the fourth industrial revolution, yet urban design also needs to consider urgently problems brought about by the rise of information technology. 'Digital Lens' has given us a deeper understanding of the construction mechanism of urban form and made 'City Portrait' more scientific. However, there are also existing personal information leaks and out of control, as the contemporary philosopher Deleuze calls it the phenomenon of 'divisions' subdivision in the 'controlled society' brought about by the code digital language, which requires vigilance, investigation and monitoring in real-time way.

The long-time contradiction between urban development and historical protection

Cities have been facing the dual challenges of renewal and decay, development and preservation. The general development trend of urban design for historical cities, historical blocks and sections is to protect and transform historical buildings and insert new ones through 'weaving and patching' on the premise of striving to preserve the original historical structure of the old city and the integrity of sections, thus forming an orderly progression of metabolism. Today, digital urban design provides a new technical tool and approach for urban historical preservation.

For example, through the digital stacking technology of historical maps, the accurate transmission of heritage historical information of different scales can already be realized. Besides, the technical problems of the overall protection and performance improvement for historical blocks and architectural heritage have been solved in a brand-new way, aiming to truly realize a win-win situation between urban development and protection.

Urban governance capacity and health

As urban development enters the new normal, it is quite necessary to deal with problems left over from the rapid development of the city in the early stage. Therefore, the 'ecological restoration' and 'urban restoration' work are carried out in a refined and sustainable way. By combining digital achievements to clarify the healthy growth of urban form and control the bottom line of urban governance, the urban development mode, 'urban disease' and urban governance capability can be effectively changed, controlled and improved.

Urban sudden disaster events

The construction of urban sanitation facilities, building's water and sewage systems, and spaces necessary to deal with sudden disasters (e.g., COVID-19) must be put on the important agenda, which is the most important core of modern urban design concepts. When unprepared disasters strike, the existing norms may be incomplete and imperfect. Special consideration should be applied to public buildings and environmental safety with important attribution to high health and safety sensitivity, high population density and possible long-term detention, such as free trade market buildings, office buildings, high-density residential communities, etc.

Combined with the big data platform, digital urban environment and health preference information are formed to provide reference for the revision, improvement and update of urban design specifications, thus effectively dealing with urban emergencies with high uncertainty.

OPTIMAL DESIGN OF URBAN-BUILDING PHYSICAL ENVIRONMENT WITH MULTIPLE SCALES

The optimal design of multi-scale physical environment of cities and buildings (including wind, heat and humidity, air and noise pollution, etc.) provides a decisive role in ensuring the safety, health and comfort of people. In the existing research of building physics, a series of numerical simulation tools based on computational fluid dynamics have been developed for building design, which can predict air velocity, temperature and pollutant distributions for small-scale indoor and outdoor environments with reasonable accuracy.

However, when applying the above simulation tools to urban planning or urban design, the following problems exist, e.g., the calculation is too slow and time-consuming to be directly applied to engineering design; the boundary interfaces of different scales of physical environment simulation are not clearly constructed; and it is difficult to validate the simulation results of real complex physical environment for large scales. Considering the above shortcomings, the integration of GIS, WRF and artificial intelligence technologies based on fast fluid prediction theories (e.g., Fast Fluid Dynamics, FFD; low-dimensional, reduced-order and linear models, etc.) is an effective way to achieve efficient and accurate simulation and design of urban-scale physical environment. In addition, the current physical environment design of cities and buildings mostly adopts the idea of pure physical environment optimization without considering the limiting factors of urban planning/urban design/architectural design, which makes the optimal physical environment design unable to be adopted in practical projects. Therefore, taking into account the design factors of urban and architectural scales and further establishing an interactive physical environment design method can promote the application, popularization and rapid development of physical environmental optimization design for urban and architectural scales.

SCALING OF SUSTAINABLE URBAN FORM

The idea of applying the concept of sustainable development to urban and architectural form has increasingly been addressed and discussed by researchers, planners and architects during the last decade. The issue of sustainable urban form has, however, both in discourse and practice, been problematic, resulting in different and contradictory results, e.g. in discussions for and against the compact city. The research results are thus often uncertain, but various guidelines and directives have been adopted and produced in recent years, often in favour of a compact city solution.

Spatial Scale

Spatial scales are continuously produced by different components, human actors, rules, built structures, objects, paving, etc. An investigation of urban form in terms scale could, for example, follow the roles of a certain urban block in different settings, acknowledging and tracking the impact of the block in systems of different sizes, levels, lengths, etc. The role of the block would of course be totally different if analysed from the scale suggested by the city grid or from the scale suggested by the perspective of a single apartment. Similarly, the spatial impact of the urban block is heterogeneous, producing certain effects at the scale of the city grid, and other effects (setting limits and opportunities) at the scale of the apartment.

The Scale of the Urban Landscape

Although it is not always fully recognized as a subject of its own, scale has always been one of the main issues of urbanity and the urban form debate. French architect and theorist Phillippe Boudon even regards it as the key concept of an autonomous architectural science, a field he refers to as architecturology. Today "regional" scale seems to be an issue of growing importance in a much larger context than that of architectural research. The transformation of towns and cities to urban regional landscapes was ongoing throughout the 20th century, starting with the trends of suburbanization, garden cities, etc. early in the century. Lewis Mumford commented in the 1930s that motorways and railroads enabled a non-hierarchical region where:

"no single centre will, like the metropolis of old, become the focal point of all regional advantages: on the contrary the 'whole region' becomes open for settlement".

Although this transformation of the urban structure was noted early, it did not become thoroughly conceptualized until the 1990s (partly owing to a vast number of influential hierarchical conceptualizations, e.g. Christaller's central place theory, The Chicago School ring model, The Athens-charter zoning systems and Newman's Defensible Space). Today, however, we are witnessing a seemingly neverending conceptual production discussing these urban transformations on the scale of the region in terms such as "Zwischenstadt", "Netzstadt", "citta diffusa", "l'urbanisme des reseaux"), the network city, the edge city, the regional city and splintering urbanism. These conceptualizations, and to some extent also mappings of urban landscapes, networks and nebulae imply that regional scale is rapidly becoming an issue of growing importance, even play with the idea of seeing the whole of Europe as one urban region, echoing Doxiades' notion of a world city, "Ecumenopolis". New regional developments, infrastructures and politics also affect and involve everyday life.

People commute more and longer, tourism is an ever-growing industry, and new institutions are being established on new scale levels. Aldo Rossi suggested as early as the 1960s that the new urban scale of the metropolitan area was more or less defined by the size of the local labour market. In the province of Skane (where Malmo is situated) the number of estimated labour markets decreased from 16 to 4 between 1970 and 2000, whereas there was, and still is, a steady increase in the number of commuters. Spatial planning is thus facing a new context in which distance has more to do with time than kilometres, and in which urbanization, greenbelts, investments, centres, etc. seem to be facing scalar shifts.

The change of perspective from city to urban landscape does not just involve a scalar shift. It increases scalar complexity as it adds a new scale of relevance to the urbanization process. The urban life of today seems to be enacted at more different spatial levels or scales than ever before. The transformation and addition of new relevant scales have, however, not yet been sufficiently dealt with in Swedish planning, where spatial development still is very much treated as territorial issues set within the frames of a "hard-edged container", which does not take into account the fact that many relations, problems and phenomena are being increasingly enacted on new and multiple scales.

Although the scale of the urban landscape is increasingly conceptualized and discussed by researchers as an empirical phenomenon, it has not yet achieved its full potential, one could argue, in more normative terms, in texts such as planning visions and in expressions of new ideas about how to build. It has even less raised the meta-question of new scalar approaches to urban issues. To some extent this is also a result of how planning practice is organized, with responsibilities hierarchically divided on different scales, often focused on intra-territorial issues and different fields of interest. These organizations can at worst result in the optimization of isolated elements, areas or aspects, which could make it more difficult to cope with the multiple relations of the urban landscapes (cf. Healy, 2005). These issues are part and parcel of the tendencies of scale stabilization discussed below, but first we need to address the question of sustainable urban form.

NEW CONCEPTIONS OF AND APPROACHES TO THE SPATIAL SCALING OF URBAN FORM

The increased pressure on cities has lead to a stronger need to build sustainable cities that last. Designing sustainable cities of the future, educated by the lessons of the past and anticipating the challenges of the future, entails articulating a multi-scalar vision and following key principles-energy, ecology, infrastructure, waste, water, livability, mobility, accessibility, economy, and culture-while responding to macro-shifts along the way. These principles are at the core of urban sustainability, which represents an ideal outcome in the sum of all the goals of development planning, on which there is widespread consensus with trade-offs and conflicts when it comes to making decisions. Indeed, research in the field of sustainable urban form, especially compact cities and eco-cities, has, over the last two decades, produced contradictory, uncertain, weak, non-conclusive, and questionable results.

The overall outcome of this research relates mostly to the actual benefits and effects claimed to be delivered by the design strategies adopted in the planning of sustainable cities. In a nutshell, the issue of sustainable urban form has, both in discourse and practice, been problematic. Much of what we know about sustainable cities to date has been gleaned from studies that are characterized by data scarcity and employ traditional data collection and analysis methods with inherent limitations, biases, and constraints, often as a result of relying on selective samples.

It is not an easy task to judge whether or not a certain sustainable urban form is actually sustainable, irrespective of the spatial scale at which such form may be considered. To some extent, the problem relates to the dilemma of form and function or structure and process, and the way this dichotomy has been conceived and approached, i.e., set up a relationship between cause and effect. New urbanism "is by necessity a fully planned and regulated environment, fiercely resistant to change and any deviation from the rigid rules that govern its form and function. But it is precisely this inflexibility, which is so important in its struggle for completion as a development enterprise".

However, Neuman argues that the form of the city is "both the structure that shapes process and the structure that emerges from a process." It follows that if form "is an outcome of evolution", then the arrangement of how to undertake planning in ways that support and guide such an evolutionary process becomes a key issue. This implies reversing the focus on urban forms governed by static planning due to its inherent limitations in achieving the goals of sustainability. Researcher argues for open, indeterminate planning due to its advantages, namely, cultural diversity; tolerance and value of topographic, social, and economic discontinuities; citizen participation; and continuous adaptation, which is common to human settlements like all other living organisms and systems.

In addition, the stable relationships between a set of sustainable activities and a certain urban form are not easily generalizable on the basis of form-function. It is widely acknowledged that the integration and balancing of the dimensions of sustainability is conflicting and contradictory, as the different aspects of sustainability rely on the different criteria for desirable outcomes. Consequently, planners will in the upcoming years "confront deep-seated conflicts among economic, social, and environmental interests that cannot be wished away through admittedly appealing images of a community in harmony with nature." Such conflicts also involve spatial interests. Focusing on the urban scale, researcher sheds more light on tendencies toward scale stabilization, i.e., the tendencies of planning from the perspective of only one or a few pre-fixed scales. The same endeavor to apply sustainable development to urban form might increase one aspect of sustainability (e.g., environmental) on one scale (e.g., the urban) while decreasing it on another (e.g., neighborhood).

There is an increasing interest in urban form as the spatial concretization of urban sustainability. At the core of sustainable urban form is the spatial pattern of the various forms of the physical objects characterizing the built-up areas of settlements at different spatial scales. The spatial is key to generating the benefits of sustainability as to its tripartite value. The existing models of sustainable urban form, particularly compact cities and eco-cities, have positive outcomes as to contributing to the three goals of sustainability.

In recent years, it is of crucial importance to find more effective ways to address and implement the spatial scaling of sustainable urban form in an attempt to increase the positive outcomes of sustainability. This relates to the emerging model of sustainable cities, which is increasingly being enabled by urban computing and intelligence in terms of planning and design as well as operational management under what has been termed "data-driven smart sustainable cities of the future".

In doing so, it highlights the innovative potential of urban computing and intelligence for enhancing and transforming the spatial scaling of sustainable urban form. This section expands on prior work done to develop a novel model for data-driven smart sustainable cities of the future in the form of a strategic roadmap towards transformational change. This new model of urbanism is grounded in the four case studies that were conducted on the prevailing paradigms of sustainable urbanism and the emerging paradigms of smart urbanism, namely:

- Compact cities;
- Smart eco-cities;
- Data-driven smart cities; and
- Environmentally data-driven smart sustainable cities.

Theoretical Background

Urban Form

It is useful to operationalize the term "urban form" as it is not easy to define. Urban form refers to the physical characteristics that make up the built-up areas in a city, including the shape, size, design, and configuration of settlements. The aggregation of these more or less repetitive elements represents urban patterns, which give a regular form to the physical objects in a city.

Researcher defines urban form as "the spatial pattern of the large, inert, permanent physical objects in a city." Urban patterns are made up largely of a limited number of relatively undifferentiated types of elements. Thus, they have strong similarities and can be grouped into concepts that comprise such components as building density and scale, block size and shape, street design and pattern, lot configuration, public space, and park layout. Urban form can be considered at different spatial scales, including regional, metropolitan, urban, municipal, neighbourhood, street, and block.

Spatial Scale

The concept of spatial scale relates to urban geography and architectural discourses. In geography, spatial scale classifies, with large approximation, the size of a land area. It is the extent of an area at which a phenomenon or a process occurs. According to researcher, scale is both the order of magnitude of a land area that shapes process and the order of magnitude of a land area that emerges from a process. The working definition in this section is based on Caniggia and Maffei's Architectural Composition that defines scale as "different level of complexity of the components internally arranged to construct a whole". The scales considered in the ambit of data-driven smart sustainable cities of the future are to be addressed as relationships between spaces of different dimensions, where the "constructed whole" of one scale can be a mere "component among components" at another. Researcher describe scales as "materially real frames of social action" and socially constructed and materially produced, and also conceives of scales as "the spatial resolution of contradictory social forces."

Sustainable Urban Form

The form of the city is seen as a salient factor for enacting more sustainable, efficient, equitable, and livable urban environments though compact and ecological design. It was the widespread diffusion of sustainable development in the early 1990s that gave a major stimulus to the question regarding the contribution that certain urban forms as human settlements might make to sustainability. Sustainable development continues to stimulate the discussion and provoke thoughts about the form of the city in light of the mounting challenges facing the world and the societal transformations triggered by the new global trends at play today. In Achieving Sustainable Urban Form, researchers conclude that sustainable urban forms are "characterized by compactness (in various forms), mix of uses and interconnected street layouts, supported by strong public transport networks, environmental controls and high standards of urban management" This characterization implies more or less a combination of the dimensions of compact cities and eco-cities.

However, management tends to dominate within the eco-city, unlike the compact city where design is at the core of compaction strategies. That is to say, the eco-city is about how the urban landscape is organized and steered rather than the spatial pattern of the characteristic physical objects in the city. Still, these two models of sustainable urban form share several concepts, ideas, and visions. Furthermore, urban form is associated with the development and design strategies related to the dimensions of urbanization, namely physical (land use change), geographical (population), economic (agglomeration), and societal (social and cultural change). Such strategies are at the core of compact cities and eco-cities.

There is no definite definition of the compact city in the literature, despite the general consensus on its common dimensions. The so-called compact city is taken to mean "a relatively high-density, mixed-use city, based on an efficient public transport system and dimensions that encourage walking and cycling." According to other views, the compact city is characterized by high-density and mixed land use with no sprawl.

Most of the existing definitions of the compact city tend to be associated with the wider socio-cultural context in which the compact city model is embedded in the form of projects and initiatives and related objectives, requirements, resources, and capabilities This also applies to the eco-city in terms of the practices and strategies adopted to achieve the goals of environmental sustainability. The idea of the eco-city is widely varied in conceptualization and operationalization. An eco-city is "an urban environmental system in which input (of resources) and output (of waste) are minimized".

According to researchers, the eco-city "is an umbrella metaphor that encompasses a wide range of urban-ecological proposals that aim to achieve urban sustainability.

These approaches propose a wide range of environmental, social, and institutional policies that are directed to managing urban spaces to achieve sustainability."

Researcher argues that there is no single accepted definition of the eco-city, but more a collection of ideas about concepts. Researcher substantiates the conceptual diversity and plurality of the initiatives and projects using the term across the globe.

Data-Driven Smart Sustainable Cities

In the context of this study, smart sustainable cities as an integrated and holistic model of urbanism is approached from the perspective of combining and integrating the strengths of sustainable cities and smart cities and harnessing the synergies of their strategies and solutions in ways that enable sustainable cities to improve and advance their contribution to the goals of sustainability on the basis of the innovative data-driven technologies and solutions offered by smart cities. Researchers specify the set of targets that need to be reached in order to attain the status of data-driven smart sustainable cities of the future (Table 1). These targets are based on the synergistic integration of the strategies and solutions of the aforementioned paradigms of urbanism. (Bibri 2021b) provides a comprehensive state-of-the-art literature review of the flourishing field of data-driven smart sustainable cities, including the definitions, dimensions, and strategies of the underlying paradigms of urbanism, namely sustainable cities, eco-cities, compact cities, smart cities, data-driven cities, and smart sustainable cities.

- Increased compactness of urban space
- High density and diversity of buildings
- Multidimensional mixed uses: social mix, physical land use mix, economic mix, and temporal mix
- Prioritized sustainable transportation and its integration with smart transportation
- Multifunctional green infrastructure for ecosystem services and biodiversity
- Balanced mixture of low-energy, energy-efficient, and passive buildings
- Large-scale net-zero and locally produced solar energy houses
- Sustainable energy system and its integration with smart energy system
- Sustainable waste system and its integration with smart waste system
- High degree of the readiness of the city to the integration of advanced technology in its management:
    - High availability and development level of the infrastructure and big data analytics competencies required for the functioning of the city

- New and extensive sources of data and a high level of support for open and standard data

- High degree of the implementation of applied technology solutions for the city management:
  - High level of the development of applied data-driven solutions for the city operational management and development planning related to the various areas of sustainability
  - Established data-oriented competences pertaining to research, innovation, strategic planning and policy, education, and professional training.

A data-driven smart sustainable city is a city that is increasingly composed of and monitored by ICT of pervasive and ubiquitous computing and thus has the ability to use the IoT and big data technologies to generate, process, analyze, and harness urban data for the purpose of creating deeper insights that can be leveraged to make decisions that accurately address the problems, issues, and challenges related to sustainability and urbanization. The emerging data-driven solutions can be adopted by city management agencies and city planning and policy centers to improve sustainability, efficiency, resilience, equity, and the quality of life. Underlying data-driven smart sustainable cities is a number of platforms and centers associated with technical and institutional competences and practices, namely:

- Horizontal information systems
- Operations centers and dashboards
- Research and innovation centers
- Educational centers and training programs
- Strategic planning and policy centers

These competences relate to the degree of the readiness of the city to introduce data-driven technology in its management as well as to the degree of the implementation of applied technology solutions in its management. The degree of readiness is characterized by the availability and development level of the technological infrastructure and competencies needed to generate, transmit, analyze, and visualize data.

The degree of implementation demonstrates the extensive use of the applied technology solutions in city operational management and development planning in relation to the different areas of sustainability. This section is associated with some of the aspects of development planning in relation to the spatial scaling of sustainable urban form.

Urban Intelligence

Urban intelligence refers to the planning, development, deployment, implementation, and maintenance of the ecosystem of big data analytics to support the interoperability between resources and technologies-and thus the integration of urban systems, the coordination of urban domains, and the coupling of urban networks-to serve the stakeholders of the city. Its functions entail the use of big data analytics and the underlying core enabling technologies to devise more effective solutions in the form of designs and responses using advanced simulation models, optimization methods, and intelligent decision support systems. This requires urban environments to be digitally instrumented to generate the data deluge that enables real-time analysis of the operating and organizing processes of urban life-urban computing. The knowledge extracted as a result of data deluge analytics serves to optimize and enhance the operations, functions, services, designs, strategies, and policies in line with the long-term vision of sustainability thanks to urban intelligence and planning functions as an advanced form of decision support.

Urban computing and intelligence has recently attracted significant attention from academia and industry for building data-driven smart sustainable cities of the future. It represents a holistic approach to harnessing and exploiting the vast troves of big data generated in cities to improve urban forms, urban infrastructures, urban environments, and urban services, as well as urban operational management and development planning systems. As such, it can generate deep insights that can be used to make well-informed and fact-based decisions, and can also create feedback loops between humans and their activities and the urban environment.

The case study is a descriptive qualitative approach that is used as a tool to study specific characteristics of a complex phenomenon. The descriptive case study approach, was identified as the most suitable approach for the empirical basis of the futures study. In this context, it involves the description, analysis, and interpretation of the four urban phenomena, with a particular focus on the prevailing conditions pertaining to plans, projects, and achievements.

That is, how the six cities selected behave as to what has been realized, the ongoing implementation of plans, and the execution of projects based on the corresponding practices and strategies for sustainable development and technological development.

To obtain a detailed form of knowledge in this regard, a five-step process tailored to each of the four case studies was adopted .

COMPACT CITIES

- Using a narrative framework that focuses on the compact city model and its contribution to the three goals of sustainability as a real-world problem and that provides essential facts about it, including relevant background information

- Introducing the reader to key concepts, strategies, practices, and policies relevant to the problem under investigation
- Discussing benefits, conflicts, and contentions relevant to the problem under investigation
- Explaining the actual solutions in terms of plans, the processes of implementing them, and the expected outcomes.
- Offering an analysis and evaluation of the chosen solutions and related issues, including strengths, weaknesses, tradeoffs, and lessons learned.

Eco-Cities

- Using a narrative framework that focuses on the eco-city as a real-world problem and provides essential facts about it, including relevant background information
- Introducing the reader to key concepts, models, and design strategies relevant to the problem under investigation
- Discussing benefits and research gaps and issues relevant to the problem under investigation
- Explaining the actual solutions in terms of plans, the processes of implementing them, and the expected outcomes
- Offering an analysis and evaluation of the chosen solutions and related issues, including strengths, weaknesses, tradeoffs, and lessons learned.

Data-Driven Smart Cities

- Using a narrative framework that focuses on the data-driven smart city as a real-world problem and provides essential facts about it, including relevant background information
- Introducing the reader to key concepts, technologies, and data-driven smart sustainable urbanism processes and practices relevant to the problem under investigation
- Providing an overview of the literature review previously conducted in relation to the study, which delivers a comprehensive, state-of-the-art review on the sustainability and unsustainability of smart cities in relation to big data technologies, analytics, and application in terms of the underlying foundations and assumptions, research problems and debates, opportunities and benefits, technological developments, emerging trends, future practices, and challenges and open issues
- Explaining the actual solutions in terms of plans, the processes of implementing them, and the expected outcomes

- Offering an analysis and evaluation of the chosen solutions and related issues, including strengths, weaknesses, tradeoffs, and lessons learned.

Environmentally Data-Driven Smart Sustainable Cities

- Using a narrative framework that focuses on data-driven smart solutions and their role and potential in improving and advancing environmental sustainability in the framework of the smart sustainable city as a real-world problem, and provides essential facts about it, including relevant background information.
- Introducing the reader to key concepts, core enabling technologies, infrastructures, landscapes, frameworks, as well as urban operating systems and urban operations centers, all with relevance to the problem under study.
- Identifying the commonalities and differences between the two cities with respect to the emerging technologies
- Explaining the actual solutions in terms of plans and visions, the processes of implementing them, and the realized and expected outcomes
- Offering an analysis and evaluation of the relevant solutions and related issues, including strengths, weaknesses, and lessons learned.

The four case studies investigated contemporary real-world phenomena with the objective to inform the theory and practice of data-driven smart sustainable urbanism by illustrating what has worked well, what needs to be improved, and how this can be done in the era of big data and in the face of urbanization. They were particularly useful for understanding how different elements fit together and (co-)produce the observed impacts in a particular urban context based on a set of intertwined factors. Overall, the four case studies conducted were useful for illuminating the different phenomena of urbanism, for illustrating the general principles underlying these phenomena, and for generating new ideas and research questions involving the relationships between these phenomena.

Field advancements in health and the built environment

As a society, our urban health challenges, or at least our understanding of them, are manifestly different in 2020 compared to the turn of the century, when the Health Map was produced. Notably, media attention to the environment and health surged after the onset of the Covid-19 pandemic with coverage including health inequities; the origins of planning, public health and sanitary infrastructure; and links between the coronavirus and climate crises. Putting the global pandemic to one side, other advancements in theory, empirical knowledge, public awareness and wider societal changes prompted the author to develop a revised healthy development framework. Three broad areas of new knowledge and theory require greater representation:

- The structural factors that determine individual health as opposed to individual characteristics and 'lifestyle' choices,
- The increased urgency of the climate crisis and other environmental breakdown, and
- The detrimental health impacts of poorly regulated development, particularly for underrepresented groups.

STRUCTURAL BARRIERS TO HEALTH

Recent debates in public health and epidemiology argue that 'lifestyle choices' are no longer a valid way to conceptualise health risk factors. Individuals may not choose to be inactive or maintain an unhealthy diet, but societal structures dictate these circumstances. The ecological models in Hancock, Dahlgren and Whitehead and Barton and Grant put significant attention on societal structures, rejecting the reductionist biomedical model that focuses on the role of individual genetic characteristics in determining health.

Despite influential critiques and advancements of other perspectives the biomedical perspective persists today. So too does the notion of unhealthy 'lifestyles', when factors of equity and inclusion should be more prominent. Researchers ecosocial theory highlights the layering of disadvantage that can affect health throughout a person's life, from exposure to pollutants during pregnancy through to poverty and race discrimination.

Krieger confronted the gaps in dominant social epidemiology theories by asking about ill health: 'do the causes lie in bad genes?, bad behaviours? Or accumulations of bad living and working conditions born of egregious social policies, past and present?'. Her ecosocial theory would point to an amalgam of these explanations through its constructs of embodiment, cumulative interplay, accountability and agency.

Krieger combines biological, ecological and social analyses to explain the determinants of health, arguing 'we literally incorporate, biologically, the world around us, a world in which we simultaneously are but one biological species among many - and one whose labour and ideas literally have transformed the face of this earth' . The ecological concepts that influenced ecosocial theory, and contemporary 'eco-epidemiology' and 'social-ecological systems perspective' theories include five core concepts: scale, level of organisation, dynamic states, mathematical modelling and understanding unique phenomena in relation to general processes. Researchers insights from Lynch and the human ecologists, Krieger's theory brings additional explanatory power for the complex interplays between people and the environment, particularly highlighting structural barriers to health.

The holistic thinking in Krieger's theory has been matched by new definitions of health and new requirements for public policies. In line with this argument, researchers use social practice theory to challenge the common public policy narrative that non-communicable diseases will be solved through individual behaviour interventions. Instead, they promote consideration of the wider structural factors influencing drinking, smoking, physical inactivity and healthy eating, including those related to the built environment such as promotion of active travel. Their position is to improve policy responses in line with an ecological health model and to push policy-makers beyond rhetoric toward action that involves confronting the vested and powerful interests that determine the main risk factors of chronic diseases and inequalities.

These new perspectives throw into question ecological health models' central positioning of individuals (and their genetic and constitutional factors) in diagrams of the determinants of health. Putting people at the centre of a model about health is logical, yet it necessarily focuses attention on individual factors as the principal determinant of health. It is argued here that built environment professionals require greater awareness of individuals' position within society and the natural environment. The language of 'lifestyles' found in healthy planning and design guidance is outdated and requires reframing. Urban development stakeholders need to aim higher to remedy unfair differences in access to health-promoting resources at multiple scales, including accessible transport, clean air and quality housing to name a few.

URGENCY OF ENVIRONMENTAL BREAKDOWN

The second area of new knowledge and theory that should underpin a revised framework for built environment professionals relates to the significant health impacts of environmental breakdown. Research on the human health impacts of environmental degradation started in the mid-19th century and now devotes significant focus to the climate crisis and other environmental breakdown. The effects of climate change are not only a problem for the future but they are also already happening and they threaten half a century of global development and health gains. This empirical evidence may only go so far in shifting public opinion (and subsequent policy action) about the causes and risks of environmental change, which is significantly influenced by wider factors including beliefs, attitudes, ideology, personal experience and other factors.

Public perception of climate change risks is rapidly changing, highlighting the increased urgency and support for policy action. The majority of respondents to an international YouGov survey in 28 countries believed that climate change would likely lead to serious economic damage, the loss of cities due to sea-level rise, and mass migration; and Asian and Pacific country respondents felt it could likely lead to extinction of the human race. Even in the USA, which previously lagged behind in public acceptance of anthropogenic climate change, there is majority support for mitigation policies, such as renewable energy.

The extensive bushfires across Australia in the summer months spanning 2019 and early 2020 caused a national public reflection on the extreme impacts of the climate crisis and associated government action. The combination of growing scientific and public understanding of climate change's health impacts for current and future populations demands a stronger reflection of climate science in built environment health guidance. In other words, decisions taken about the built environment will affect human health for local and distant populations over time as a result of the distributed impacts of the climate emergency.

URBAN DEVELOPMENTS' IMPACT ON UNDER-REPRESENTED GROUPS

A final area of field advancements relates to increasing knowledge about the health impacts of different forms of urban development and inequitable impacts across specific groups in society. There is increased recognition that market-driven and poorly regulated development can have detrimental health effects, and yet there has been an over-reliance on this form of growth in many countries to meet housing demand and fund urban improvements.

Researcher describes a shift in British planning beginning in the 1970s where reductions in public sector funding were accompanied by a de-valuing of planning's regulatory role in protecting the public interest. Although these trends play out differently across planning systems, they have generally reduced planners' ability to leverage positive health outcomes from development.

An adversarial relationship has been documented between planners and developers in negotiations about how urban development will support local health. For example, an Australian planner working near Melbourne described being 'beholden' to developers to increase the local housing supply.

The planner recalled a case where they requested improvements to a development, including pedestrian-friendly design, and the developer threatened to abandon the project entirely. Gaps in public sector funding to improve ageing infrastructure or meet local housing demand may be filled by the private sector where a commercial return can be gained, but a growing body of literature raises concerns about the health impacts of these new developments or regenerated areas.

These challenges relate to well-known dynamics of power, inequity and the under-representation of groups who are most affected by urban development. Although these issues have been simmering for decades, the epidemiological evidence base has recently grown stronger and given weight to the measurable health impacts of such developments.

On top of growing health-related evidence, other societal shifts have pushed new aspects of equity and inclusion to the forefront of healthy urbanism debates. Scholars and practitioners are increasingly exposing the unfair (health) impacts of urban development caused by failure to design for residents' diverse needs with regard to gender, age, race and other characteristics. Criado-Perez's book exposes the impact felt by women when built environment design and management strategies do not account for their needs. For example, she highlights researchers research on the mismatch between women's safety needs on public transit and the strategies adopted by transit operators. The implications of this mismatch are clearly linked to women's physical and mental health - if they are deterred from transport they may be less active (and therefore at higher risk for chronic diseases) and have reduced access to employment opportunities. Similarly, Researchers documented the bicycle infrastructure needs of low-income minority communities in Boston and found their requirements to be absent from transport and Crime Prevention Through Environmental Design (CPTED) guidelines, representing the likely ubiquity of exclusive mobility infrastructure. Beyond a narrow focus on disabilities, recognition that some groups in society need different design features to live a healthy life is a recent shift in built environment theory and practice.

The proposed THRIVES Framework responds to the advancements in theory, science and practice outlined above and the previously described gaps in existing frameworks. In the author's view, there is an urgent need to reframe health and its wider determinants for built environment audiences. There is a need to more fully integrate the complex interconnections across sustainability, equity and inclusion concepts to demonstrate how urban policy and development affect health locally and remotely in temporal and spatial terms. The next section introduces the THRIVES Framework with and explanation of its component parts in terms of theory and practical implications.

THE THRIVES FRAMEWORK

The THRIVES (Towards Healthy uRbanism: InclusiVe Equitable Sustainable) Framework offers a new way of conceptualising the interconnected health impacts of built environment policies and design decisions at multiple urban scales. The Framework can be used to inform research and practice in the fields of urban planning, architecture, urban design, engineering, transport, public health and others.

The visual representation of the Framework uses words and images to evoke processes and outcomes of healthy urbanism. The visual illustrations show the planet at the centre which is connected to regional, peri-urban and urban landscapes moving outward. The illustrated spatial scales align to the three scales of health impact (planetary, ecosystem and local). Concentric rings in the image demonstrate interconnections across each scale and help to visually align the scales of health impact with scales of urban decision-making related to the built environment. The latter begin with regional (in recognition of cities' role in their wider geographical region) and then move to city, district, neighbourhood and building scales. Decision-making scales will vary across cities and forms of development. The Frameworks seeks to represent scales that are indicative of policy or decision-making for typical government tiers or types of urban development. The scales of health impact and decision-making are associated with evidence-based design and planning goals. All decision-making should be informed by three core principles that define what healthy urbanism 'looks like': inclusive, equitable and sustainable.

The Framework aims to communicate three broad propositions. First, there are complex interactions between urban environments and health that result in health impacts at multiple spatial and temporal scales that may not be immediately obvious to designers and policy-makers. Interactions across scales are visually represented in THRIVES through the nested circles and the arrows intersecting the illustrations. Researchers summarise key characteristics of complex systems as they relate to healthy urbanism. Complex systems are dynamic, comprised of many elements, interconnected and governed by feedback.

Such systems contain non-linear structure and result in emergent behaviour and counterintuitive results. Achieving healthy urbanism requires systems thinking approaches, integrated design and interdisciplinary working. Applying a systems approach will result in many co-benefits (or win-win-win solutions) for social, economic and environmental goals. Second, knowledge to inform decisions about health and urban environments should come from multiple sources, including scientific (and other technical) evidence and locally affected communities. This will ensure that urban development better serves groups which have historically been under-represented in design, such as women, minorities and children.

Finally, urban environments and their associated health and wellbeing impacts should be monitored by government, building managers and other responsible authorities (again using participatory processes to define indicators) to ensure that policy and design intentions result in health-promoting places for all residents. The following sections describe the Framework, explaining how urban policy and design decisions will affect peoples' health and wellbeing through diverse mechanisms, resulting in impacts that may be spatially or temporally distant to where decisions are made.

THREE CORE PRINCIPLES

To the extent that it is possible, all design and policy decisions should be inclusive, equitable and sustainable. In the author's experience, these three concepts are viewed as independent goals by built environment professionals and they are often the remit of specialist consultants. A holistic understanding of the interconnections between these principles is lacking, resulting in a narrow view of how the urban environment affects health. This section makes the argument that each principle is essential to healthy urbanism by providing definitions, relations to health and underlying theoretical concepts.

Inclusive

Inclusive urban environments represent both an outcome to strive for and a process to follow. Built environment professionals have varied understandings of inclusive design and how it should be applied, leading to limited adoption of this principle to date. Furthermore, developers and property owners have been reluctant to respond to statutory accessibility requirements, often citing cost as the principal barrier. The topic is historically rooted in designing environments that include people of all physical abilities and ages. Even this narrow definition put significant responsibility on design professionals as Clarkson and Coleman argued that people are not disabled by 'physical and mental impediments' but 'by designs and environments that do not take account of the full range of human capabilities'. Global trends demand inclusive design as the ageing population and those with chronic conditions are more likely to have physical impairments that require built environment design responses such as sloped level changes and more places to rest in public spaces, among other measures. The concept of inclusion has broadened over time and it now includes factors such as gender and race. Design for all means 'design for human diversity, social inclusion and equality'. As argued in the previous section, the built environment design needs of women and minorities have been largely ignored, resulting in potential negative health impacts.

In the THRIVES Framework, an inclusive built environment enables all members of society to conveniently participate in daily activities without feeling that they are disadvantaged by their personal characteristics or needs, and this is achieved through participatory processes. This definition builds on existing definitions and highlights a subjective element because the emotions associated with feeling excluded are also damaging for health and wellbeing. Individual characteristics that require design considerations will vary across projects and internationally. A useful starting point for characteristics to consider are those protected under the UK's Equality Act 2010: age, disability, gender reassignment, marriage and civil partnership, pregnancy and maternity, race, religion or belief, sex and sexual orientation. In addition, built environment design and policy should consider the requirements of people with:

physical and mental health conditions that are not classified as disabilities, such as declining cognitive function associated with dementia; people with several of these characteristics who may feel particularly excluded; and other vulnerable people such as homeless or low-income communities.

Healthy urbanism requires the participation of all members of society in design and planning processes. Researchers claimed 'that development cannot be sustainable and long-lasting unless people's participation is made central to the development process'. The need to integrate community knowledge in design and planning to promote health was articulated through the transport examples from Loukaitou-Sideris and Fink's research on women and public transit and Lusk's work on the cycling infrastructure needs of low-income minority communities. Actively seeking community knowledge about health and place is important because their perspectives may differ to those of design professionals. Inclusive urban development processes build on the participation, co-production and co-design principles of health and urban design and planning. Groups that will be affected by urban policies and designs should participate in deliberations over what is proposed and how it will be achieved. Furthermore, citizen groups may take the lead through various self-build and community-led development approaches.

Equitable

The broad concept of equity is about fair distribution of resources, but THRIVES focuses on health equity, or the absence of unfair differences in health that are caused by uneven distribution of health-promoting resources in society. Researcher argues that achieving health equity requires 'striving for the highest possible standard of health for all people and giving special attention to the needs of those at greatest risk of poor health, based on social conditions'. Income inequalities reduce population health and wellbeing and people living in poorer neighbourhoods tend to die earlier and spend more time in ill health than people living in more affluent neighbourhoods - known as the social gradient in health.

The relations between income inequalities and health are not part of core educational curricula for most built environment professionals. Therefore, these practitioners are also unlikely to draw connections between poverty and other social and environmental inequalities that influence health and wellbeing in cities, such as reduced access to healthy housing, green infrastructure and employment.

There is a strong evidence base outlining the health impacts of reduced access to health-promoting environmental resources. Residents in low-income or ethnic minority neighbourhoods have reduced access to healthy foods and greater access to unhealthy foods. Low-income communities are more likely than affluent communities to have increased exposure to air pollution, noise pollution and rates of road traffic injuries.

They are more likely to live in poor quality housing that is overcrowded and less able to withstand extreme heat, cold, flooding, earthquakes and natural disasters). Deprived communities also have reduced access to green space, fewer recreation facilities, and reduced availability of children's play areas.

All of these factors are directly influenced by built environment decisions and they should therefore be a core consideration in urban planning and development.

In the THRIVES Framework, an equitable environment gives access to health-promoting environments to all residents and specifically considers and seeks to reduce barriers to access (be they physical, cultural, social or economic). The public health principle of 'proportionate universalism' offers an approach to reduce inequity that can be applied to urban policy and development.

Adopting Carey et al.'s approach to proportionate universalism for healthy urban environments would involve measures that are universal (applied to everybody in society, such as safe water) and targeted (applied to those with the most need, such as high-quality social housing). It would also mean adopting the governance principle of 'subsidiarity', where residents are closely involved in determining solutions to the challenges they face (as argued under the inclusion principle above). There are overlaps between the concepts of inclusion and equity. For the purposes of the THRIVES Framework, equity is about overcoming unfair distribution of resources, whilst inclusion is about ensuring everybody can fully participate in society and daily activities, achieved through a process of participatory design and planning.

SUSTAINABLE

The final core principle in the THRIVES Framework is sustainability. The term sustainability and the commonly referenced 'triple bottom line' balancing of economic, environmental and social objectives are contested and complex. Researcher argues that continued economic development is not compatible with the limited resources of the planet. However, it may still be required in many parts of the world, particularly low-income settings that rely on development to reduce poverty and improve quality of life. High-income countries, such as the UK and Australia, have also relied on growth to fund community benefits through new development and the success and long-term viability of this has been called into question.

The UN Sustainable Development Goals (SDGs) (United Nations General Assembly 2015) provide a foundation for combining urban health and sustainability agendas from global to local scales. The SDGs attribute greater importance to both environmental limits and the role of urban policy than previous global sustainable development initiatives and they are aligned to health equity.

The THRIVES Framework considers sustainable development to be supportive of the needs of the current (and immediately local) population without compromising the needs of future (or spatially distant) populations, building on the Brundtland Report definition. The Framework builds upon Agyeman's concept of 'just sustainabilities', simplified to 'just sustainability' by Rydin. This broad view of sustainability is about process and outcome, and it relates to improving wellbeing, equity and justice for current and future generations, and doing so within ecosystem limits. Researcher emphasises that these principles are mutually supportive, stating that 'both social and environmental health are dependent, to a large extent, on greater justice and equality'. The principles of just sustainabilities are articulated through both the three core principles and the three scales of health impact in the THRIVES Framework.

THREE SCALES OF HEALTH IMPACT

There are three interconnected scales of health impact: planetary health, ecosystem health and local health. In contrast to other frameworks that start with individual characteristics, THRIVES positions core principles and planetary health at its core, shifting the focus from individuals to our environment. Impacts across these scales are not all immediate and may occur over months, years or decades, within and beyond the boundaries of physical environment changes.

Built environment decisions often result in changes that last decades or longer, and it is therefore essential to consider how they will impact health at multiple spatial and temporal scales. The author argues that measuring the health impact of urban policy and development (e.g. through health impact assessment or healthy building rating tools) is typically limited to consideration of effects that are spatially and temporally proximate to the proposed change. To create healthy places, professionals must consider impacts at broader scales. It is not necessary to model such effects for all projects - researchers have already created the evidence base that can inform policy and design decisions. This evidence is described below and grouped into three broad scales that create a design heuristic to inform practice.

Planetary health

The concept of planetary health refers to 'the health of human civilisation and the state of the natural systems on which it depends'. The built environment causes environmental degradation at a global scale through resource-intensive design, construction and operation and through habitat destruction that reduces biodiversity.

Unplanned rapid urbanisation and deforestation also results in people living closer to 'untouched ecosystems' where there is greater risk for transfer of zoonotic diseases. Local and global environmental harms affect human health in multiple interrelated ways and the impacts can be felt in distant locations to such harms. Environmental pollution has often affected poor populations disproportionately to richer groups and this inequity plays out at multiple scales within cities and internationally. Low-income groups have contributed the least carbon emissions (and other harms) but will suffer the most from the health impacts of climate change, such as malnutrition and death or injury from extreme weather.

The built environment can support planetary health through three goals (intended as key examples): enhancing biodiversity and promoting resource efficiency and zero carbon. At the urban scale, the most significant design and planning decisions that affect planetary health are made at the regional to city policy levels. Researchers describe how health is supported by reducing carbon emissions through land use and transport planning. Reducing motor vehicle travel supports climate change mitigation, reduces injuries and supports active travel. In turn, reducing traffic pollution leads to reduced noncommunicable diseases and increased social capital which improves mental health and wellbeing. Decreasing traffic will also support more inclusive cities, recognising that road injuries are the biggest cause of death in people aged 5 to 29. Other scales of urban decision-making are relevant to support planetary health (all of the scales are interconnected). For example, striving for a zero carbon built environment would mean increasing energy efficiency in buildings which can improve thermal comfort, thus demonstrating the interconnections across multiple health impact scales.

Ecosystem health

Ecosystem health as a concept relates more to the health of the environment than the connection between the environment and human health; however, the THRIVES Framework emphasises the latter. Ecosystems are 'webs of connections between living and non-living system components' and they are foundational to human health. As with planetary health, there is a complex system of interconnections between ecosystem and human health. Under an ecosystem services approach, ecosystems provide food, clean water, climate regulation and recreational opportunities, and in turn, these services support human health and wellbeing. Similarly to planetary health, harms to ecosystem health are felt most closely by low-income populations who depend heavily on their local environment, such as subsistence farmers.

The THRIVES Framework lists five example goals for ecosystem health: sustaining air, water and soil quality, and greenspace, and improving sanitation, waste, and mobility infrastructure. Highly impactful decisions that affect ecosystem health are made at the scales of regional to district policy and design - a point illustrated through the design and planning goal of water quality. Water is a necessity for human and ecosystem health, and urban growth increases water demand, which will be exacerbated by climate change. One in four global cities is water stressed, increasing the importance of sustainable water management, such as through avoiding development in flood-prone areas, using permeable landscape surfaces, green roofs and other landscaping to slow down water drainage and installing rainwater harvesting systems to collect and use water. Improving sanitation and waste infrastructure will protect water quality and reduce exposure to infectious diseases.

Local health

At the neighbourhood scale, example design and planning goals are to connect people with services (covering employment, education, retail, leisure, healthcare and other facilities), perceived and actual safety, culture, public space and food. At the building scale, example goals are to shelter people with acoustic and thermal comfort, affordability, tenure security, lighting and space. There are many pathways through which different types of buildings (e.g. offices, homes, schools, hospitals, etc.) and neighbourhood design (e.g. compact or sprawling urban form) influence health. A large body of literature and local knowledge (from professionals and community groups) can be used to further prioritise or complement the example design and planning goals in THRIVES.

As with planetary and ecosystem health, this scale interacts heavily across the others. A healthy building or neighbourhood must support human health alongside ecosystem and planetary health, as argued by Levin. For instance, creating walkable neighbourhood environments will support physical activity and social connection whilst reducing local air pollution and greenhouse gas emissions. Walkability is achieved through decisions taken at multiple scales including urban land-use policies, mobility infrastructure, location of services and quality of public space.

In most cases, the author argues that health and sustainability are mutually reinforcing design and planning goals, however, there may be some tensions that bare consideration. At the building scale, increased energy efficiency could result in overheating or poor air quality, conversely efforts to filter out air pollutants could increase energy use in buildings. Integrated design and planning should be adopted at all scales to avoid these tensions wherever possible. The next section highlights the potential co-benefits of healthy urbanism by examining two urban development scenarios at neighbourhood and building scales.

MOVING FROM THEORY TO PRAXIS

A primary goal of the THRIVES Framework is to reframe current perspectives on how the built environment affects health and wellbeing, ultimately shifting practice in urban planning, design and development. Several hypothetical development scenarios are used here to illustrate practical application of the Framework. A brief summary of the participatory workshop feedback follows, outlining potential paths to implementation.

The process of moving from the conceptual framing of the THRIVES Framework to actual policy and design decisions is explained through two hypothetical urban development scenarios of different scales and types. Each scenario is explored through a single design strategy that produces health benefits across the three core principles and scales of health impact. These benefits are not achieved automatically and the table is not comprehensive in assessing health impacts.

Design strategies need to respond to the local context and be informed by community knowledge. In the first scenario, a local authority has received funding to improve mobility infrastructure in a deprived neighbourhood. They achieve this goal by installing new solar-powered street lighting, pedestrian crossings, cycle lanes and street furniture (such as benches and planters). In the second scenario, a commercial developer has partnered with a local authority to refurbish and expand a 1960s housing estate with a mixture of market-rate and affordable properties. Their approach involves upgrading energy efficiency and ventilation in dwellings, providing publicly accessible outdoor play space, and installing new sustainable drainage and ground-floor storage for cycles and strollers.

Professionals who attended the participatory workshop gave several ideas for how the THRIVES Framework could be used to inform practice. The participants included a mix of built environment and public health professionals working in the public and private sector in England. Participants identified the Framework's potential impact in changing professionals' perspectives on how urban development impacts health, expanding existing conceptualisations to a broader and more holistic set of topics. Professionals thought the Framework could break down disciplinary barriers and help to build a shared understanding of healthy urbanism.

Finally, they listed multiple practical uses of THRIVES, including informing local planning policy, supporting impact assessments (health, environmental, integrated, etc.), and aligning development proposals to specific health goals.

Moving from theory to practice with the THRIVES Framework will require more than the publication of a high-level conceptual diagram. An advisory group has been convened to guide the co-development of a short training programme with relevant professionals. The roles of different actors will vary when using THRIVES and this requires further exploration in terms of their diverse perceptions and requirements. To that end, the THRIVES Framework is being tested through implementation in the design and planning process of a large-scale urban development in London. It is hoped that the training programme and additional insights gained through using THRIVES will lead to greater understanding of the mechanisms that are successful at improving health through urban development over time. In this way, the author proposes that the Framework may inform research about the process and outcomes of healthy urban development.

CONCLUSION

The predicted growth in urban areas demonstrates the significant opportunity that built environment professionals have to shape future physical and natural infrastructure in a way that will positively impact people and the planet. The THRIVES Framework can support this task. As researcher claims 'human thriving depends upon planetary thriving', yet not all in society agree to work within the planetary and human wellbeing boundaries articulated in her doughnut framework. In response to this, THRIVES could be useful to support interdisciplinary collaboration and build shared understanding among practitioners working on design, policy development or impact assessment. Recommendations for professionals arising from this research are to use THRIVES as the basis for discussions with partners or within teams. These conversations could begin with participants sharing their views about how their work affects the different scales of health impact and the three core principles. The Framework could also be used to conduct a mapping exercise of how design or policy strategies affect health. These activities may reveal gaps, but they may also show that professionals were unknowingly doing work that supports health.

The proposed THRIVES Framework fills gaps in previous conceptualisations and responds to field advancements across diverse sectors in science and society. The novel contributions of THRIVES are its assemblage and visual portrayal of diverse theory and empirical evidence about health and the urban environment, aimed at (and tested with) a built environment audience, specifically the following three insights. First, THRIVES highlights the structural factors that determine health and de-emphasises individual 'lifestyle' choices. Second, it proposes greater attention to the complex links between local, ecosystem and planetary health that result in health effects that are both proximate and distant in space and time to urban development. Third, it explicates the interconnected principles of equity, inclusion and sustainability that are not typically understood as key requirements for health. The concepts described in this section are too often siloed, diluting the arguments and resources to build and maintain healthy places. The THRIVES Framework offers a holistic conceptualisation of healthy urban environments that can build shared understanding among researchers and practitioners.

BALANCING THE SCALES OF CITY SUSTAINABILITY

Because people, goods, and services - and, significantly, environmental media such as air, water and energy - cross jurisdictional boundaries all the time, what happens in one part of a region can profoundly affect what happens in another part. We cannot have sustainable cities and towns if their surrounding regions are not sustainable.

Second, even if a region as a whole has the empirical markings of sustainability, it will not be a place in which many of us want to live if its component neighborhoods are not also healthy, hospitable and attractive. Simply put, neighborhoods are where we encounter our environment on an everyday basis. They are the places that matter the most to us, and that affect our health and well-being most directly.

Neighborhoods are also where increments of change place, and thus where we must pay attention and where we can most easily make a difference. This goes for commercial and mixed-use neighborhoods as well as for residential districts.

If metro regions represent the economic scale of people habitat, neighborhoods represent the human scale. We have to get them both right.

Growth challenges ahead

All sorts of neighborhoods - and all sorts of regions - are going to be facing pressure over the next several decades. This is because the overwhelming majority of American metro areas - collectively, where some 85 percent of us live - are going to be experiencing substantial growth. While the expansion will occur in some parts of the country more dramatically than in others - the Phoenix and Tucson metros are projected to more than double in population from 2010 to 2050, while the Great Lakes region is projected to add "only" about 28 percent to its population in that period - the challenges faced across the country in accommodating growth will differ mainly as a matter of degree.

By now, we know all too well how not to do it: In the late 20th century and, in many places, continuing into the 21st, we allowed population, jobs, and investment dollars to sprawl across the countryside, quite literally in search of greener pastures on which to build as we allowed our older neighborhoods and towns to deteriorate. The result of this rampant spreading out was desecration of the landscape as farms, forests, and watersheds were paved over; massive growth in motor vehicle traffic and accompanying pollution as we had to drive longer and longer distances to get things done; increased strain on municipal budgets as infrastructure and service dollars were spread too thin; and economic and social stress for populations left behind, all too often in pockets of concentrated poverty.

In order to have any chance at achieving sustainability in the face of growth, we must do things differently. We must use land more efficiently, building whenever we can at walkable densities on vacant and underutilized land within the existing development footprint. We must put more people within convenient reach of jobs, shops and amenities and close to comfortable and reliable transportation options.

Where our older communities and neighborhoods have been weakened, we must strengthen them. Where our newer suburbs have been built haphazardly, we must take advantage of opportunities to revisit their configuration and make them work better. Where our natural and rural environment has been threatened, we must take care to conserve it. (Same with the built historic fabric of our cities and towns, by the way.) Where constituencies have been left out of planning and development decisions in the past, we must include them. These are the values that underlie what some of us call urbanism, what others of us call quality growth or smart growth. Urbanism is generally very helpful to the environment at both the regional and the neighborhood level.

At the regional scale, urbanism is a terrific asset to sustainable growth management; more compact development patterns create a range of efficiencies that lower the per-capita impacts of regional growth in comparison to sprawl. At the neighborhood scale, walkability brings considerable public health benefits and obviates what might otherwise be motor vehicle trips.

Urban and walkable suburban living is popular in the marketplace, too; demand for urban living is growing. Indeed, a new analysis of real estate data published in The Washington Post shows that city and inner suburban neighborhoods across the country are rising in value much faster than their outer suburban counterparts.

But it is also a bit more complicated at the neighborhood scale: unless mitigated and managed very carefully, some of the local environmental consequences of urbanism can actually be negative rather than positive.

The geography of impacts from traffic

Compare, for example, the two color-coded maps of metropolitan Philadelphia, below. The one on the right shows greenhouse gas emissions per capita from transportation. The green areas on the map indicate the parts of the region where per-capita emissions are the lowest; they are concentrated in Center City Philadelphia, inner suburbs, and other urban centers sprinkled around the region. This is consistent with research showing that people who live in relatively central and relatively dense locations do not drive as far or as often, on average, as people who live in more sprawling locations, shown on the map in red.

If you are a regional planner looking to reduce growth in traffic and associated emissions, you want to concentrate as much new growth as possible in those green locations where per-capita impacts are relatively low, and to retrofit the yellow and red locations with less sprawling, more walkable and, yes, denser development so that their performance improves over time. But take a look also at the map on the left. This one shows not emissions per household but emissions per acre, and in this case the more urban places are the ones in red. This map is telling us where environmental impacts from traffic are greatest and that density, notwithstanding its per-capita advantages, can bring more traffic and along with it more tailpipe emissions to some places. If you are a resident of a neighborhood destined for more urban, denser development, and you are fearful that the proposed development will bring more congestion and more pollution, the map on the left is suggesting that your fears may not be irrational.

Similar issues for watersheds

There is a similar dynamic going on with respect to runoff water pollution from development. In particular, the volume of stormwater runoff flowing into receiving waters during a rain event tends to vary with the amount of impervious surface (mostly pavement and rooftops) in the watershed: the greater the amount of impervious surface, the more pollution. This is because impervious surfaces prevent water from filtering into the ground gradually, forcing it instead to "run off," picking up contaminants along the way.

Just as with motor vehicle traffic, development density lowers per-capita impervious surface, and thus runoff volume, in the watershed compared to large-lot sprawl. While this may seem counterintuitive because large lots might be thought to have more lawn and forested area and thus more absorbent ground cover, the additional perviousness within each individual lot is not enough to overcome the large amount of impervious surface imposed by the longer, wider roadways and bigger surface parking lots associated with sprawling development. Faced with a growth scenario, an environmentally conscious regional planner thinking about watershed health might want to encourage compact development and discourage sprawl.

But, again, the wrinkle is that density, unless mitigated, tends to increase local, per-acre imperviousness. Here is another set of maps, this time from Seattle:

Note how the per-capita imperviousness is greatest in the outlying, more sprawling parts of the region, while the per-acre imperviousness is greatest in the central, more densely developed parts. The latter is one of the reasons why urban waterways tend to be more polluted than rural ones, and one of the reasons while restoring the health of our nation's urban waterways (and their often-overloaded sewer systems) can be so challenging.

So, where does this take us?

If you are looking for easy answers to resolving this fundamental dilemma - what I call in my book "the environmental paradox of smart growth" - you will not find them here. The hardest challenge a planner or other responsible official attempting to manage growth might face is to strike an appropriate balance between sometimes legitimately competing concerns, and the challenge must be met with collaboration, empathy, sophistication, and creativity. I might even argue that finding the right balance in the right situations is what the city planning profession is for.

We do have tools with which to work: With regard to motor vehicle traffic, we can take steps to mitigate potential increases in traffic volume and its associated harms (which include noise as well as congestion and air pollution) with increased coordination of development with non-motorized transport modes such as walkability, bicycle facilities, and good transit service. Researcher find it hopeful that recent research suggests that traffic generated in urbanizing locations is not nearly as bad as traditional engineering theory might suggest. And there are models to emulate: Indeed, in my own region Arlington, Virginia has had great success adding considerable density to a transit corridor without the great increases in traffic that some predicted. More generally, the Washington, DC metro region as a whole grew in population by 14 percent between 2007 and 2015 while increasing motor vehicle traffic by only one percent.

With regard to stormwater pollution, researcher is very encouraged by the rapidly evolving field of green infrastructure: by using techniques such as strategically designed tree plantings and landscaping, green roofs, rain gardens, pervious pavement, and rain barrels, we can capture stormwater even in highly urban settings and allow it to filter into the ground before it becomes polluted runoff and contaminates receiving waterways. Pocket parks and community gardens can be part of the solution, too. What I love most of all about green infrastructure is that in addition to its water quality benefits it brings nature and badly needed green space into dense urban neighborhoods, making them simultaneously more sustainable and more lovable.

Researchers also think that moderation can be part of the answer. Intensive development has its place; but, as researchers have written before, researchers love the idea of walkable neighborhoods that retain a human scale, and researchers believe the concept holds a lot of promise for bringing us the best of both worlds in the right situations. We do not need to duplicate Manhattan or Hong Kong densities to show substantial improvements in environmental performance over sprawl.

On this point, researchers find it encouraging that there is a growing movement in architectural circles to promote a smaller-scale, more incremental approach to urbanism. The idea is that real estate development in the US has come to be dominated by Big Projects backed by Big Financing; and smaller-scale, more incremental approaches that can strengthen neighborhoods without overwhelming them have become rare. Thought leaders such as Dan Parolek and John Anderson are working to turn that around, quietly building a constituency and a national portfolio of successful smaller projects, spreading the word in workshops around the country. Think small-scale infill such as two- and three-story, multi-family buildings often found side-by-side with single-family homes in older neighborhoods across the country.

RELATIONSHIP BETWEEN SUSTAINABLE URBANISATION AND URBAN RENEWAL

Majority of the world's population now live in cities, and the number of people living in the countryside is steadily decreasing every day. Because of the population growth and unplanned settlements in urban centres, the available resources are running out and natural habitats are undergoing changes. In other words, failure to meet the growing need for housing in urban areas leads to unplanned settlements in a way that threatens both the identity and liveability of the city. In addition, this type of construction has brought with it the land zoning and infrastructure problems. This process has also accelerated the destruction of green areas, agricultural areas and coastal areas. And the places, where unhealthy, dilapidated and illegal buildings are intense, create visual pollution as well as become a potential threat to the health of the people living in them and the fabric of cities.

Urban spaces should not be considered only as a space consisting of buildings, but also as a social and cultural centre where social events take place. Therefore, due to the significant impacts of urban areas on the society and human behaviours, their design should also include the social and cultural aspects in addition to the physical dimension. As there is not a particular perception of space, it is considered as a concept of place that is constantly changing and renewed together with those living in it.

When looked at the urban transformation projects in the world, cities seem to have different regeneration processes from each other according to their own internal dynamics due to the impact of globalisation. Intervention strategies of the developed and developing countries include different contents from each other, and the cities that have completed their regeneration eventually force the other cities into this process. So, local governments attaching importance to the implementation of urban transformation want to create quality urban environments, renew the city's image and raise their status and position. For example, Canada, Norway, Switzerland, Sweden and Australia are among the countries that apply the ecosystem successfully in the urban transformation projects, the protection of natural resources and the use of renewable energy sources. In this respect, today Sweden is a model country with its capital Stockholm, chosen as the ecological capital of Europe in 2010.

URBAN RENEWAL

According to Zheng urban generation, urban renewal, urban redevelopment and urban rehabilitation share same meanings in the town-planning field, but they are considerably different in term of scale. Urban renewal and urban regeneration express nearly the same meaning and both involve work of a relatively large scale. Urban renewal is defined as a process of slum clearance and physical redevelopment taking account of other elements such as heritage preservation; while urban regeneration is a comprehensive integration of vision and action aimed at resolving the multi-faceted problems of deprived urban areas in order to improve their economic, physical, social as well as environmental conditions. There are also many other striking definitions, explanations and application forms of this concept.

In the dictionary of the Turkish Language Institution, the term 'renewal' is defined as transformation, reform, to enter into another form and to take another case. When called 'urban renewal' in Turkey in the 1970s and 1980s, the first thing that came to mind was the improvement of slum areas. Actually, urban renewal is a general concept including the applications carried out for the renewal of the existing urban structure.

Generally, urban renewal has been regarded as a sound approach to promoting land values, improving environmental quality, rectifying the urban decay problem and meeting various socioeconomic objectives on the one hand, and enhancing the existing social networks, improving the inclusion of vulnerable groups and changing adverse impacts on the living environment on the other. Urban renewal is also considered as a comprehensive work aiming at improving the physical, social, economic and ecological aspects of urban areas through various actions including redevelopment, rehabilitation and heritage preservation.

According to researchers, urban renewal projects can profoundly improve urban competitiveness providing that they are designed and implemented properly. On the other hand, urban renewal is associated with other concepts such as sustainability and defined in a broader sense. For example, according to Czischke, sustainable urban renewal is understood as renewal actions, policies and processes within a city, which addresses interrelated technical, spatial and socio-economic problems in order to reduce environmental impact, mitigate environmental risk, and improve environmental quality of urban systems, lifestyles and assets.

Historical development of urban renewal process

The renewal of the post-war era that began in the 1950s with rebuilding to repair the destruction, continued as revamping in the 1960s and 1970s. Then, a new process was launched in the in the 1980s and 1990s although they are different in every country, and especially the projects dominated by the private sector and corporate partnerships were implemented. Today, urban renewal activities including all of the concepts such as urban regeneration, urban conservation and urban gentrification are performed all over the world in order to ensure the sustainability of the city. Unfortunately, many urban renewal work carried out especially in developing countries is far from sustainable urbanisation and urban ecological understanding, for the reasons such as the lack or national policy and local policy, lack of infrastructure, urban rents fight and so on.

Sustainability, sustainable development and sustainable urbanism

Today there is a widespread belief that different dimensions of 'sustainable development', such as economic, social, environmental and institutional, are not given equal priority by many policy makers. But it is not considered as a rational or comprehensive application form. Researchers, argue that sustainability is a concept including social economic and ecological aspects. In addition, UN Brundtland Commission dated 1987, suggests sustainability is a development that meets the needs of the present without compromising the ability of future generations. Researchers define sustainability as a significant concept which not only aims at high productivity but also intends to continue the functions of any social, economic or ecological system requiring sustained continuity without damaging or depleting the sources used.

On the other hand, the term 'sustainable development' dates back to the 1970s and it is a relatively complex concept made even more complicated by the fact that there is still no commonly accepted definition of this term. Sustainable development makes the world a better place to live in, for both the present and future generations. It also involves the preservation protection of the earth's wealth-creating sources by bringing about the social and economic conditions for a transformation in that direction. Besides being the determinant of environmental policy in the world, sustainable development is also defined as a concept integrated with the concepts of economic and social development without getting limited to the domain of the environment.

According to Berke and Manta, sustainable development is a dynamic process connecting local and global concerns, in addition to linking local, social, economic and ecological issues in order to fairly meet the needs of current and future generations. Corresponding to urban renewal in terms of social, economic and environmental sustainability, sustainable development has been recognized that urban renewal and sustainability should be combined.

Although sustainability and sustainable development are not the same, they are often used without attention as if they were. According to Barrow, sustainability is the ongoing function of an ecosystem or use of a resource, and implies steady demands. But sustainable development implies increasing demands for improving well-being and lifestyles and also in the foreseeable future, for a growing population.

As for sustainable urbanism, it is a widely used phrase, often with ecological and green connotations, constituting a rather complete framework for the interdisciplinary planning and urban design of contemporary cities, neighbourhoods and residential places. In addition, Hall emphasized that 'Planning and renewal must not be separated; instead, renewal must be an integral part of planning'. This approach explores sustainability and urban design in a holistic manner by focusing on the processes that shape the form and function of our built environment in its full complexity

THE LINKS BETWEEN URBAN RENEWAL AND SUSTAINABILITY

Especially in the early developed cities, urban areas grow and deteriorate as time goes by, and adversely changing conditions of the environment make life miserable for the people living in them. So urban renewal projects to improve the built environment take place. Emphasizing the importance of sustainability in this context, researchers argue that 'the battle for sustainability will be won or lost'.

In general, urban land is both precious and scarce, like some other sources. Thus, the main objective of urban planning is providing reuse of the already scarce urban sources taking into account the public interest. To do so, urban planning should be obliged to enable the public interest principle, and establish and apply necessary rules reconciling private interests with the public interest depending on the character of the economic structure and political regime of every country.

Nevertheless, decisions on the future of cities should be taken with the participation of the relevant stakeholders mainly consisting of the state sector, local government sector, private sector, civil society organisations, local residents as well as academics. In this context, a sustainable urban renewal strategy should take into consideration the entire life cycle of urban structures from design and construction to the operation and maintenance as well as life quality of the residents.

Finally, as a living organism, the city is a hybrid system consisting of structures with natural areas. But, rapid and unplanned urbanisation is the biggest threat in terms of the sustainability of the natural and cultural sites. According to Sancar, technical infrastructure and transportation are extremely important for sustainable organisation and also decisive in terms of the quality of urban life. The adequacy or quality of the technical infrastructure is considered to be among the issues to be addressed first. In addition, transportation systems of the vehicles and pedestrian in the city need to be re-evaluated within the understanding of a holistic planning approach in order to achieve the goal of sustainable urbanisation.

CHAOPTER

TECHNOLOGY ESSENTIALS 9

Contemporary cities have a pivotal role in strategic sustainable development; therefore, they have gained a central position in operationalizing this notion and applying this discourse. This is clearly reflected in the Sustainable Development Goal 11 (SGD 11) of the United Nations' 2030 Agenda, which seek to make cities more sustainable, resilient, inclusive, and safe. Sustainable cities have long been the leading global paradigm of urbanism for more than three decades or so. The subject of "sustainable cities" remains endlessly fascinating and enticing, as there are numerous actors involved in the academic and practical aspects of the endeavor, including engineers and architects, green technologists, built and natural environment specialists, and environmental and social scientists, and, more recently, computer scientists, data scientists, and urban scientists.

All these actors are undertaking research and developing strategies and programs to tackle the challenging elements of sustainable urbanism. This adds to the work of policymakers and political decision-makers in terms of formulating and implementing regulatory policies and devising and applying political mechanisms and governance arrangements to promote and spur innovation and monitor and maintain progress in sustainable cities.

Since its widespread diffusion in the early 1990s, sustainable development has had significant positive impacts on the planning and development of cities in terms of the different dimensions of sustainability. It has also revived the discussion about the form of cities. In this regard, it has inspired a whole generation of urban scholars and practitioners into a quest for the immense opportunities and fascinating possibilities that could be enabled and created by, and the enormous benefits that could be realized from, the planning and development of sustainable urban forms (especially compact cities and eco-cities), That is to say, forms for human settlements that can meet the required level of sustainability and enable the built environment to function in a constructive way. This can be accomplished through continuously improving their contribution to the goals of sustainable development in terms of reducing material use, lowering energy consumption, mitigating pollution, and minimizing waste, as well as in terms of improving equity, inclusion, the quality of life, and well-being.

However, new circumstances require new responses. This pertains to the spread of urbanization and the rise of ICT and how they are drastically changing sustainable urbanism. The transformative force of urbanization and ICT and the role that cities can play have far-reaching implications. By all indicators, the urban world will become largely technologized and computerized within just a few decades, and ICT as an enabling, integrative, and constitutive technology of the twenty-first century will accordingly be instrumental, if not determining, in addressing many of the conundrums posed, the issues raised, and the challenges presented by urbanization.

It is therefore of strategic value to start directing the use of emerging ICT into understanding and proactively mitigating the potential effects of urbanization, with the primary aim of tackling the many intractable and wicked problems involved in urban operational functioning, management, planning, development, and governance, especially in the context of sustainability. Indeed, the rapid urbanization of the world poses significant and unprecedented challenges associated with sustainability due to the issues engendered by urban growth. In short, the multidimensional effects of unsustainability are most likely to exacerbate with urbanization. Urban growth will jeopardize the sustainability of cities.

Therefore, ICT has come to the fore and become of crucial importance for containing the effects of urbanization and facing the challenges of sustainability, including in the context of sustainable cities which are striving to improve, advance, and maintain their contribution to the goals of sustainable development. The use of advanced ICT in sustainable cities constitutes an effective approach to decoupling the health of the city and the quality of life of citizens from the energy and material consumption and concomitant environmental risks associated with urban operations, functions, services, strategies, and policies.

Smart sustainable cities as an integrated and holistic approach to urbanism represent an instance of sustainable urban planning and development, a strategic approach to achieving the long-term goals of urban sustainability—with support of advanced technologies and their novel applications. Accordingly, achieving the status of smart sustainable cities epitomizes an instance of urban sustainability. This notion refers to a desired (normative) state in which a city strives to retain a balance of the socio-ecological systems through adopting and executing sustainable development strategies as a desired (normative) trajectory. This balance entails enhancing the physical, environmental, social, and economic systems of the city in line with sustainability over the long run—given their interdependence, synergy, and equal importance. This long-term strategic goal requires, "fostering linkages between scientific research, technological innovations, institutional practices, and policy design and planning in relevance to sustainability. It also requires a long-term vision, a trans-disciplinary approach, and a system-oriented perspective on addressing environmental, economic, social, and physical issues." All these requirements are at the core of backcasting as a scholarly and planning approach to futures studies. This approach facilitates and contributes to the development, implementation, evaluation, and improvement of models for smart sustainable cities, with a particular focus on practical interventions for integrating and improving urban systems and coordinating and coupling urban domains using cutting-edge technologies in line with the vision of sustainability.

One of the most appealing strands of research in the domain of smart sustainable urbanism is that which is concerned with futures studies. The relevance and rationale behind futures research approach are linked to the strategic planning and development associated with long-term sustainability endeavors, initiatives, or solutions. And backcasting is well suited to any multifaceted kind of planning and development process, as well as to dealing with urban sustainability problems and challenges.

The aim of this chapter is to generate a vision for smart sustainable cities of the future by answering the 6 guiding questions for step 3 of the futures study being conducted, namely:

- What are the terms of reference for the future vision?
- How does the future sustainable socio-technical system and need fulfillment look like?
- How is the future vision different from the existing socio-technical systems?
- What is the rationale for developing the future vision?
- Which sustainability problems, issues, and challenges have been dealt with by meeting the stated objectives and thus achieving the specified goals?
- Which advanced technologies and their novel applications have been used in the future vision?

TECHNOLOGICAL INNOVATION

Overview

Technological innovation involves "the introduction of something new, or a new idea, method, or device". Sustainable development must take into account the effectiveness of the technology. Technological innovation, as part of technological change, allows organisations to test new ideas at speeds and prices that were never anticipated a decade ago. "Changes in consumption patterns can drive the creation of new technologies necessary for sustainability and their adoption and diffusion at the desired pace. Success in bringing about these changes will require substantial reorganisation of the economy and society and changes in lifestyles. Economic and financial incentives for the creation and adoption of new technologies will be needed, which may include innovative policy reforms".

Innovation in technological change

Innovation is related to how consumers perceive an idea, a practice, a product, or an object as a new thing, and begins with consumers' awareness of the innovation. Meanwhile, diffusion is a process in which information about the innovation is conveyed through certain channels over time among the consumers. Researcher discussed that innovation goes beyond invention. It refers to the spread of new technology into the wider society.

The development of technological innovation has led to the sixth generation. Researcher mentioned that the development of technological innovation in the first stage up to the fifth has already begun since the 1950s. The generations of innovation include technology push, market/demand pull, coupling, cross-functioning, as well as integration and networking. Researchers added another generation, a sixth generation, known as design-driven innovation. It means to design new products whilst simultaneously creating a need for consumers. Technological change and sustainability are closely related to each other. Both factors form the innovation in order to improve the effectiveness of environmental stewardship, social development, and economic progress.

Technology is transforming innovation at its core, and the speed of change involves evolution, effect, and what is ahead. Researchers revealed three distinct technology trends that are poised to be of the highest priority in terms of facing rapidly accelerating digital innovation. The combination of digital innovation in the business model will establish and empower a city to be more sustained.

- **Transparently immersive experiences.** Technology becomes more human-centric, where it will introduce transparency between people, businesses, and things. Also, the evolution of technology becomes more adaptive, contextual, and fluid within the workplace, at home, and interacting with businesses and other people.
- **The perceptual smart machine age.** Smart technology will be the most disruptive over the next decade. This will allow organisations with smart machine technologies to harness data in order to adapt to new situations and solve problems.
- **The platform revolution.** Emerging technology is revolutionising the concepts of how platforms can be defined and used. This includes shifting technical infrastructure to ecosystem-enabling platforms, which is laying

the foundations for entirely new business models that are forming the bridge between humans and technology.

Although a research related to technology innovation has reached 100 years, the convergence of research related to science, technology, and innovation only began in 1960. The overall technological innovation system has garnered widespread attention, especially beyond the context of sustainability. Besides technological innovation system, there have been other aspects that are related to system delineation, spatial aspects, transitions, politics, and policy recommendations.

There are two main components related to the study of innovation. First is the use of patent scale, whereby the data is used to study the geography of innovation and also the spatial extent of knowledge spillover. Second is the measuring of innovation by using survey. This refers to a set of questions that asks establishments whether a new product, process, or organisational innovation has been introduced. The questions are also related to whether the innovation is new.

One of the main problems of urban sustainability is related to energy consumption. Researcher, in his study, focused on the sustainable perspectives of energy consumption. The results showed that the energy saved in 1 year could be used power 60,000 European homes. In practice, manufacturers, especially, can improve energy efficiency through design and implement energy-efficient projects on their own or enter into energy performance contracting (EPC), which mainly includes shared and guaranteed savings. Also, Miao's study discussed the overall energy system and CO_2 emission. The population scale is the most important factor in residential energy consumption and CO_2 emissions.

The world has witnessed extreme environmental deterioration over the past few decades. Rapid development in urban areas has caused severe environmental pollution and ecological destruction. Green innovation is an important approach to achieve sustainable city development. There is even a relationship between legitimacy pressure, profitability, and green innovation. Green innovation will lay the foundation for balancing profit and environmental responsibility, and at the same time achieve the harmonious development of socioeconomic environmentally. The continuity of technological innovation involves green technological innovation, green institutional innovation, and green business model innovation.

Technological innovation is not just related to road management system, building management system, and public facilities management, but also to solid waste management. The problems of solid waste management include the amount and composition of waste generated, rapid expansion of urban areas, funding issues, rapid technological advancement, and limited energy and raw materials.

To ensure the success of technological innovation, it requires a consolidation policy Researchers proposed layering and parallel policy-making. It is a complementary concept to understand the challenges associated with the implementation of sustainable mobility. A number of studies have also been identified to be related to policies such as environmental policies, transport policies, and policy and planning for shrinking cities.

SUSTAINABLE CITY

Overview

A sustainable city means "development that meets the needs of the present without compromising the ability of future generations to meet their own needs". In the twenty-first century, there has been a shift from sustainability assessment to smart-city goals, transformed into smart sustainable city.

The framework of a city should be holistically integrated with the urban sustainable principle. Basically, most of the literatures that have been studied so far correlated innovation and city development together. Innovation is considered as a driver of economic growth. Innovation is also an open process, with innovative organisations drawing ideas, contacts, and learning opportunities from their environment.

Four dimensions of sustainable development have been identified in building a more sustainable city:

- Environmental dimension is related to the objectives of the conservation and preservation of the environment. This includes aspects such as atmosphere, soil, water, oceans and seas, biodiversity, and sanitation.
- Social dimension is related to the objective that involves the satisfaction of human needs. They include population, employment and income, health, education, housing, and security, which aim to reflect the level of education, income distribution, and living conditions.
- Economic dimension is concerned with the efficiency of production processes and changes in consumption patterns.
- Institutional dimension includes the ability and effort spent by governments and societies to implement changes required for an effective implementation of sustainable development.

Environmental dimension

"Cities are at the crossroads of implementing system changes for sustainable development. Paradigm shifts are needed in multiple dimensions to redefine the relations of cities with the environment". "Cities impact their immediate environment, as well as the global environment". According to researcher, a sustainable city includes environmental sustainability, heritage conservation, appropriate technology, infrastructure efficiency, placemaking, social access, transit-oriented development, regional integration, human scale, and institutional integrity.

Energy consumption in urban areas is as much as 75%. It also accounts for 70% of carbon dioxide emissions. To meet the needs of the population in urban areas, infrastructure development sometimes ignores ecological sustainability. Living in urban areas has contributed to pressure on the ecosystem. Human ecological footprint has exceeded the capacity of terrestrial ecosystems for as much as 50%. The level of ecosystem services at the moment is risky at a critical level of degradation. Thus, there are five major challenges for urban sustainability. These include ecological footprint, ecosystem services and biodiversity, investment for sustainability, the good life, as well as leadership and cooperation.

So far, a number of current issues related to the environment have been published. There is a close relationship between sustainable urban development and environmental sustainability. Therefore, it is important to formulate any strategy of transformation, renewal, and regeneration. A few studies have even evaluated the environmental outcomes and effectiveness of regeneration programmes, especially in relation to green spaces, access to public transportation, climate change, and seismic-risk reduction.

Due to the high population density in the city, transport system is a key issue being discussed. Researchers mentioned that an efficient and flexible transport system is essential for a future of low carbon transport system. Their study also found that the differences in the levels of emissions coming from the different systems of each city depict differences in the quality and utilisation of transportation networks. Carbon emissions are correlated with income. Per capita incomes are higher in a city, which results in higher average per capita demand in major emissions sources. High carbon emissions have also increased from households. By understanding that behaviours affect energy consumption will be useful in promoting low carbon household.

Researcher discussed a green city as a new approach for planning and developing a sustainable city. Sustainable urban development appears to be closely connected with the discursive production of green city or eco-city. The implementation of sustainable urban development focuses on green infrastructure, transit-oriented development, brownfield redevelopment, and carbon reduction plan. The purpose of a green city is to strive to reduce environmental impact by reducing waste, expanding recycling, lowering emissions, and increasing housing density, while expanding open space and encouraging the development of sustainable local businesses.

Thus, a number of studies have been conducted that are related to green value. A study conducted by researcher stated that green building construction is considered to be an essential practice to achieve sustainability. Researchers discussed the importance of connecting land use and water planning. Researchers' study emphasised on flood mitigation infrastructure, whereby it is crucial to remove accumulating floodwater from low-lying areas where drainage is naturally poor due to a very low slope gradient. A study by Zhang concluded that the operating efficiency, operating performance, and service quality of a drainage enterprise is determined by its sustainable development level that involves urban wastewater treatment, construction of drainage engineering, operation and maintenance, and consulting. Researchers evaluated ecosystems and the impacts caused by the change in the environment. It involves making the allocation of land for urban development more sustainable and bears an impact on food production, ecosystem services, and biodiversity conservation.

The Economist Intelligence Unit (EIU) developed a green city index in alignment with green management. It covers CO_2 emissions, energy, buildings, land use, transport, water and sanitation, waste management, air quality, and environmental governance (refer to Table 1). Another model of green management proposed 12 green guidelines for creating a happier, healthier, and a more sustainable and green city. The 12 green guidelines fall into three key categories: urban form (urban growth boundary, transit-oriented development, mixed-use, small blocks, and public green space), transportation (non-motorised transit, public transit, and car control), and energy and resources (green buildings, renewable and distributed energy, waste management, and water efficiency). They are also consistently moving towards to a smart city, which involves smart telecommunication, smart mobility, smart energy management, smart governance, smart public services, and smart safety.

EIU'S THE GREEN CITY INDEX

Air quality

- Nitrogen dioxide
- Sulphur dioxide
- Ozone
- Particulate matter
- Clean air policies

Buildings

- Energy consumption of residential buildings
- Energy-efficient buildings standards
- Energy-efficient buildings initiatives

CO_2

- CO_2 intensity
- CO_2 emissions
- CO_2 reduction strategy

Energy

- Energy consumption
- Energy intensity
- Renewable energy consumption
- Clean and efficient energy policies

Environmental governance

- Green action plan
- Green management
- Public participation in green policy

Transport

- Use of non-car transport
- Size of non-car transport network
- Green transport promotion
- Congestion reduction policies

Waste and land use

- Municipal waste production
- Waste recycling
- Waste reduction policies
- Green land use policies

Water

- Water consumption
- System leakages
- Wastewater system treatment
- Water efficiency and treatment policies

Social dimension

The aspect of social development towards building a more sustainable city "occurs when the formal and informal [processes, systems, structures, and relationships] actively support the capacity of current and future generations to create healthy and liveable communities". This social aspect covers equity, diversity, quality of life, democracy and governance, and inter-connectedness.

Researchers addressed two major issues about social quality in culture-led urban regeneration plans. First is a plan in terms of the instrumental involvement of the community in participation practices. Second is an urban policy that is related to the creative city notion that needs to be implemented.

Living sociability involves the city itself and a combination of culture and nature. The study conducted, measured the link between cultural heritage management and sustainable urban development and identified three levels of a monitoring system of the management system: strategic level, operational level, and monitoring level.

The World Economic Forum identified three indicators for social sustainability. The first indicator aims to assess a population's access to basic necessities.

It is a measure of inclusion, as well as a measure of the fulfilment of basic physical needs.

The second indicator is linked to the concept of perceived economic security. It aims to evaluate a population's vulnerability to economic exclusion. Finally, the third indicator assesses social cohesion.

Economic dimension

There is a relationship between balance economic goals with social and environmental. As mentioned by Soubbotina, this balance can be discussed with reasons:

- If environmental and social losses resulting from economic growth turn out to be higher than economic benefits, the overall result for people's wellbeing becomes negative.
- The economic growth itself inevitably depends on its natural and social conditions. To be sustainable, it must rely on a certain amount of natural resources and services provided by nature such as pollution absorption and resource regeneration. The economic growth must be constantly nourished by the fruits of human development.

Researchers studied the economic contribution to the sustainability and efficient functioning of a city, especially the benefits in an economic context of private individual benefits and public goods benefits.

- First, the economic assessment focuses on the impact on businesses.
- Second, in terms of private benefits, transport investment provides a mixture of private and public goods benefits. The time-saving in private benefits is a key motivation.

- Third, for business sales, perceptions of change in sales have been found to vary by location and nature or characteristics of products.

Fourth, in terms of health benefits, it includes time-saving and reduced costs of congestion, pollution, and CO_2 emissions. The cycling activities provide physical exercise, which contributes to health and reduces expenditure on healthcare.

Finally, in terms of time-saving and wider economic benefits, it contributes to the city's economy by improving connectivity between journey, origin, and destination, and by reducing journey time.

THE IMPACT OF TECHNOLOGICAL INNOVATION ON BUILDING A SUSTAINABLE CITY

It is very important to understand the challenges of a contemporary city. The challenge is due to the increasingly complex nature and the variety of the contemporary city. Thus, a future city needs to balance economic development with energy efficiency and the preservation of natural resources in order to improve the quality of life. "Balancing the immediate needs of today without compromising the needs of tomorrow is at the heart of being a sustainable city, and of this report".

A study done by researchers demonstrated the nature of sustainable innovation, which is sustained by the materiality, social structures, and institutional frameworks that comprise the overall sociotechnical system in which innovation takes place. The social value can be leveraged for shaping alternative courses of action, creating technological innovation, and developing novel practices that create sustainable value for society.

Sustainability challenges have affected the formation of planning and the vision to create a future city or smart city.

- Han proposed the conceptual vision of a sustainable urban future in 2050. To achieve a low carbon society, various technological advances in transportation, buildings, and industries are among those that can contribute towards this end.
- Kourtit eproposed a roadmap for innovative strategic urban policy. This includes developing urban agglomeration as engines of creativity and technological innovation; exploiting technological opportunities and developing a sustainable urban system; laying the foundation for social cohesion and social capital; and developing ecological footprint and improve climate neutrality of urban areas.
- Future urban configurations should concentrate on the efficient use of resources and opportunities in contributing to economic growth; deploying infrastructure, physical assets, and amenities; providing social services;

minimising poverty, inequalities, and segments of the population; and protecting the environment and preserving natural assets.

Researchers released a list of the 100 most sustainable cities in the world. Most cities are from Europe and Asia. Ten of the most sustainable cities are Zurich, Singapore, Stockholm, Vienna, London, Frankfurt, Seoul, Hamburg, Prague, and Munich. Though there are three significant areas of correlation (people or social, planet or environment, and profit or economic), there are also other elements that make a city more sustainable:

- The urban regeneration and innovation hubs for the future development, as the city transforms itself from a manufacturing centre to a knowledge, technology, and innovation centre for the nation.
- The urban planning policies to strengthen the city's identity, global competitiveness, development direction, and innovation in the living environment.
- Adopting technological innovation, especially new innovative technologies, cyclical processes, and business models.

FROM SMART BUILDINGS TO COGNITIVE CITIES

The digital evolution of cities has been underway for several years. It has created a virtuous circle that resulted in cities becoming increasingly competitive, sustainable and prosperous through the optimization of urban activities and infrastructure and, at the same time, accelerated innovation and the creation of new opportunities.

The road ahead is still a long one, though, as can be seen from the manifest ferment in the industrial sector. In fact, according to a recent McKinsey & Company study, about $10 trillion will be spent to create smart factories, buildings and infrastructure in the world each year. This expenditure will increase to $14 trillion by 2025 and will represent 13% of the turnover of the entire construction industry, which will make possible an enormous increase in the productivity of this sector.

Fig. 1.

Clearly, the private sector alone cannot embark on this path to innovation. Close collaboration with the public sector is needed to accelerate a comprehensive innovation process that can also include the general public.

THE EVOLUTION OF SMART CITIES

The evolution of smart cities has been a gradual one. We started from an initial technology-driven perspective with small but fundamental changes - for example, the synchronization of traffic lights. We then moved on to a technology-enabled and city-led viewpoint in which technologies such as the Internet of Things (IoT), artificial intelligence (AI) and the cloud created cities we know today, with a range of digital services available.

A further step forward is the citizen co-creation of the cities of the future. In this paradigm, citizens have a fundamental role in guiding a radical transformation of cities according to their needs and goals, but also their desires and dreams. Technology and humanism combine to create the cities of tomorrow, with individuals who commit themselves to create, develop and test every type of innovation born to meet their needs.

The next step is the creation of cognitive cities: New entities that are different from smart cities, where there's a continuous process of interaction and exchange between technology and residents and in which solutions such as AI are crucial in these ongoing improvements.

From Smart Buildings To Cognitive Cities

A fundamental factor in the creation of smart cities is smart buildings, the request for which continues to grow thanks to technologies such as the cloud that guarantees more effective and accessible digital twins, big data, cognitive service and machine learning services, IoT that provides enhanced sensor connectivity and standardized, open protocols, as well as, of course, AI. These technologies are increasingly available and within the range of investment.

However, the creation of smart buildings is also guided by the demand for reduced management costs, the possibility of offering new and unusual experiences and, above all, the need to create more productive workplaces. In fact, we imagine a future in which an office building gathers information about meetings for the following day and combines this data with the weather forecast and the preferences of workers in terms of the use of the meeting rooms while allowing for optimization of energy consumption as part of the growing awareness around sustainability.

The Emergence of The First Cognitive Cities

Across the globe, there are numerous examples of successful application of cognitive city technologies.

My company, Microsoft, has moved in this direction and adopted cutting-edge technologies to make its Redmond headquarters smart and connected. Its campus - comprising 125 buildings, 2,000,000 data points that register 500,000,000 transactions every 24 hours - required total digitalization to become a single asset, thus reducing waste and improving building efficiency. Thanks to the introduction of technologies equipped with AI, we generated enormous benefits, including energy savings of 6% to 10%, while improving technical assistance (over 10,000 requests / month) and correct 48% of problems in just 60 seconds. And all with an ROI achieved in just 18 months.

We also followed the Yas Island leisure parks developed by Miral, a United Arab Emirates company that creates tourist attractions throughout the world. With the goal of attracting 48 million visitors by 2022 and understanding the importance of having a complete, in-depth overview of visitor behavior, Miral decided to invest in digitalization of its spaces to guarantee personalized, immersive and totally new experiences for 100% involvement of every visitor.

PRACTICAL ADVICE FOR SUCCESS

Our learning from building successful cognitive buildings and cities is to focus on inclusive digital infrastructures that integrate new technologies into the existing context. Successful plans rest on five key principles:

- **Connect top and bottom services.** Consider the way "base services" (power, air, water, etc.) interact with middle (access, signage, security, occupancy, etc.) and top services (productivity, commerce, etc.).
- **Adapt to specific contexts.** Adopt new technologies, both when returning to the office post-pandemic and when designing new spaces, leveraging at best the existing context.
- **Always be inclusive.** As real people live in buildings, carefully consider the impact of technology on them, considering generational, gender and cultural differences.
- **Take a living view.** You should always consider the way that buildings and cities are "lived" in different ways during different parts of the day and night.
- **Involve society at large.** Seek creative contributions from both public and private entities which may be impacted and/or benefit from the cognitive building.

A LOOK INTO THE FUTURE

Naturally, by definition, all technologies we have spoken about are in constant evolution and, because of them, citizens can experience the city of the future on a daily basis, thanks to the use of spaces in ways that are necessarily very different than in the past. We have spoken of citizens, but digital experiences will impact all players in the urban universe, whether visitors, tourists, tenants, workers or maintenance personnel.

They will be the ones impacted by the experiences that can be created when technology, social aspects and the way of living these spaces are melded into one. Some of these concepts already exist, others are in the process of being developed and it's only by continuing to invest in this sector that we will be able to amaze ourselves with future applications.

TECHNOLOGIES THAT ARE ESSENTIAL FOR BUILDING A SMART CITY

When people think of smart city technologies, the words they're most likely to think of are 'IoT' and 'sensors'. And this association is not unwarranted or unfounded, as by far the biggest enabler of smart cities will be their intricate network of interlinked sensors collecting all kinds of data from citizens, infrastructure, and pretty much everything else. However, while advanced sensor technology will play a critical role in empowering smart cities, a smart city is reliant on many more technologies other than sensors. These data-driven technologies function together in order to optimize resource utilization, enable convenience for citizens, and provide greater visibility and control to governments. Following are the different technologies that will form the core of smart city infrastructures:

When people think of smart city technologies, the words they're most likely to think of are 'IoT' and 'sensors'. And this association is not unwarranted or unfounded, as by far the biggest enabler of smart cities will be their intricate network of interlinked sensors collecting all kinds of data from citizens, infrastructure, and pretty much everything else. However, while advanced sensor technology will play a critical role in empowering smart cities, a smart city is reliant on many more technologies other than sensors. These data-driven technologies function together in order to optimize resource utilization, enable convenience for citizens, and provide greater visibility and control to governments. Following are the different technologies that will form the core of smart city infrastructures:

The Internet of Things

All the other technologies in this list have always been part - albeit in more rudimentary forms - of government operations. However, none is as much associated with the concept of smart cities as IoT is. That's because smart cities, and also smart governance for that matter, rely on the collection, analysis, and processing of large volumes of granular, real-time data, which has only become possible with the help of IoT sensors. IoT sensors and cameras can constantly collect detailed information in various forms and in real time. Data like the footfall at a railway station, traffic on a road, contamination levels in a water source, and the energy consumption in a residential block can be collected in real time using different types of IoT sensors. By using this data, government agencies can make quick decisions pertaining to the allocation of different resources and assets.

For instance, based on the footfall and ticket sales information from railway stations, transport agencies can reroute their trains to cater to the varying demand. Similarly, health, safety, and environmental agencies can monitor the contamination levels of water bodies and notify the people responsible to take remedial action. In certain cases, IoT actuators can automatically initiate responsive actions in exigent situations, such as stopping the supply of contaminated water to domestic households.

Thus, IoT networks and sensors will essentially form the nervous system of smart cities, delivering critical information to the controlling entities and relaying responsive commands to the the appropriate endpoints.

Big data analytics

All aspects of smart cities will be primarily driven by data. All the decisions - from long-term strategic ones like public policy to short-term decisions like evaluating the welfare worthiness of individual citizens - will be taken through the analysis of pertinent data. With the increased volume, velocity, and variety of data generated thanks to IoT sensors and other advanced data gathering methods, the need for high-capacity analytics tools will be greater than ever before.

Big data analytics tools are already being used by governments for a wide range of applications, from predicting the likelihood of crime in specific areas of the city to preventing crimes like child trafficking and abuse. As IoT enables the collection of data from a host of new sources, big data analytics will be used in all areas including critical ones like education, healthcare, and transportation. In fact, big data analytics is already an inextricable part of IoT networks, and can potentially be subsumed under IoT in the future. Big data can enable governments to make sense of the data they possess by finding patterns that are reflective of trends within cities. For instance, big data analytics can help education departments to spot trends like low retention rates so that they can prevent such outcomes. Big data can also be used to find the causes of such issues and plan remedial actions. Thus, big data will become a critical decision-making aid to smart city governments.

Artificial intelligence

Building upon the capabilities of IoT and big data will be artificial intelligence. Artificial intelligence can be used to bolster smart cities' big data and IoT initiatives by automating smart decision making. In fact, IoT's ability to initiate responsive action will be largely driven by some or the other form of artificial intelligence.

The most obvious application areas for AI in smart cities is automating high-volume and data-intensive tasks like offering basic citizen services in the form of chatbots. For instance, AI-based chatbots can be used to process simple requests like:

- Providing new water, electricity, and gas connections,
- Changing property ownership details, and
- Changing addresses and other registered details on government records.

However, the real value of artificial intelligence can be derived by smart cities by utilizing advanced AI applications like deep learning and computer vision. For instance, computer vision can be used by transportation regulators to analyze traffic footage to identify instances where drivers park illegally. Computer vision can also be used to look out for and flag vehicles that are associated with reported crimes to help law enforcement track down criminals.

Deep reinforcement learning can also be used to automatically optimize resources based on emerging needs in smart cities. With the help of reinforcement learning, governments can make their operations increasingly efficient as these AI systems can get better with experience.

5G CONNECTIVITY

Smart cities are founded upon the ability of their different subdivisions to communicate and share information in real time for ensuring complete synchronicity in operations. By achieving this synchronicity, governments can ensure that their citizens receive critical services like healthcare, emergency response, and transportation in a timely manner. As a result, they can not only ensure their citizens' convenience but also improve their safety and overall wellbeing. For instance, in case of an emergency like an explosion or a fire, enabling real-time communication between the fire department, city ambulance service, and traffic control units can ensure perfect real-time coordination among these entities, potentially keeping casualties to a minimum. To enable this seamless communication between the different government entities, it is important to have a communications network capable of handling high volumes of communication with low latency and high reliability. By using 5G communication technology, governments can ensure that all government bodies can collaborate seamlessly, despite the volume of data shared every minute.

Augmented reality

Offering citizens timely services means ensuring that the government employees are given the information they need to perform their tasks effectively. For instance, a physician at a government health centre must be given information on the patients they are working with. Or, workers tasked with fixing a damaged railway line should be given the updated layout of the tracks and the accurate location of the damaged part. By using an AR headset, such information can be relayed to the workers in real-time as soon as they need it. This minimizes the effort and time required by the workers to find the necessary information. As a result, they can initiate action without any delays.

AR can also be used by traffic management personnel to get real-time information regarding parking violations and stolen vehicles through smart glasses or smartphone applications. Thus, they can ensure that drivers have a convenient experience during their commutes.

Each of these smart city technologies, as you may have surmised already, are interdependent on each other. Thus, governments should develop an integrated strategy for implementing these technologies in conjunction with each other, taking into account the advantages brought about by each of them.

HOW IS IOT TECHNOLOGY MAKING CITIES SMARTER AND BETTER?

Secure wireless connectivity and IoT technology are transforming traditional elements of city life - like streetlights - into next-generation intelligent lighting platforms with expanded capabilities.

The scope includes integrating solar power and connecting to a cloud-based central control system that connects to other ecosystem assets.

These solutions shine far beyond simple lighting needs.

- High-power embedded LEDs alert commuters about traffic issues, provide severe weather warnings, and provide heads up when fires arise, for example.
- Streetlights can also detect free parking spaces and EV charging docks and alert drivers where to find an open spot via a mobile app. Charging might even be able from the lamppost itself in some locations!

In addition to people, dwellings, commerce, and traditional urban infrastructure, there are four essential elements necessary for thriving smart cities:

- Pervasive wireless connectivity
- Open data
- Security you can trust in
- Flexible monetization schemes

WHAT'S THE BEST WIRELESS TECHNOLOGY FOR SMART CITIES?

The first building block of any smart city application is reliable, pervasive wireless connectivity. While there's no one-size-fits-all, evolving Low Power Wide Area Network (LPWAN) technologies are well suited to most smart city applications for their cost efficiency and ubiquity.

These technologies include LTE Cat M, NB-IoT, LoRa, Bluetooth, and a few others that all contribute to the fabric of connected cities.

The advent of 5G technology is expected to be a watershed event that propels smart city technology into the mainstream and accelerates new deployments.

Fig. 2.

Opening the data vault

Historically, governments, enterprises, and individuals have held their data close to the pocket, sharing as little as possible with others.

Today, open data is redefining the digital city.

Privacy concerns and fear of security breaches have far outweighed the perceived value of sharing information (see: Portland and privacy).

However, a key enabler of sustainable smart cities is that all participants in the complex ecosystem **share information and combine it** with contextual data analyzed in real-time.

This is how informed decisions are made in real-time.

Multiple sectors must **cooperate** to achieve better, sustainable outcomes by analyzing real-time contextual information shared among sector-specific information and operational technology (OT) systems.

The conclusion?

Data management (and access to this information) represents the backbone of the digital city. Stay with us. Here is what we mean.

EXAMPLES OF SMART CITIES

- The New York City Department of Transportation's Midtown in Motion is a **congestion management system** that has improved travel times on Midtown's avenues by 10%.
- The NYCx Challenges initiative from the NYC Mayor's Office of the Chief Technology officer invites entrepreneurs, technologists, and tech professionals to participate in open competitions and propose bold ideas that solve real urban needs such as pollution, income inequality, and transport (site closed).
- LinkNYC provides **free super-fast free Wi-Fi**, phone calls, device charging, and a tablet to access city services, maps, and directions. It's a unique communications network replacing payphones across the Bronx, Brooklyn, Manhattan, Queens, and Staten Island.
- Cyber NYC is the city's **strategic investment to dominate cybersecurity**. It aims to grow New York City's cybersecurity workforce, help companies drive innovation and build networks and community spaces.
- MyNYCHA mobile app and web portal allow public housing residents to manage services online. It addresses over 300 public developments across New York City. Launched in 2015, **MyNYCHA** is a free service that puts the repair process in residents' hands. Residents can submit, schedule, and manage work tickets online. They can also subscribe to alerts for outages in their developments, view inspection appointments, and pay their rent.
- **Biking in New York City**: Plan for cycling in the city.
- **Automated water meters** in NYC: Automated Meter Reading systems consist of small devices connected to individual water meters. They send daily readings to a computerized billing system.
- The **My DEP Account** lets New Yorkers track consumption from home. The system eliminates the need for a water meter reader to visit premises. It allows the Department of Environmental Protection to monitor citywide consumption more closely and manage the city's water supply system more effectively.
- New York's data report - Open Data for All- provides free public data published by various local agencies. This tool opens data for people to make a difference in their communities—including educators, students, artists, builders, small business owners, advocates, reporters, community board members. It also means open data for the 300,000 workers who make New York City safer, cleaner, and more equitable.

Amsterdam Smart City

Amsterdam is a shining example of a well-connected smart city reaping the rewards of opening the data vault. The Smart City initiative began in 2009 and included over 170 projects.

It also shares **traffic and transportation data** with interested parties such as developers, who then create mapping apps connected to the city's transport systems.

Now, navigating the city is a snap for all.

The city built autonomous delivery boats called 'roboats' to keep things moving in a timely fashion.

It also supported a floating village of houses, solving the city's overcrowding problem with a sustainable, energy-efficient alternative. Power is generated within communities, and homes receive water straight from the river and filter it within their tanks.

Copenhagen Smart City

Copenhagen is known as one of the smartest cities in the world and mobilizes expertise worldwide.

The city is leveraging open data to collaborate with the Massachusetts Institute of Technology (MIT) to develop an **innovative intelligent bike system**.

Embedded with sensors that provide real-time information to both the riders and administrators, data is shared to monitor and manage air quality and traffic congestion.

More:

- Technologies to create data-driven solutions that suit Copenhagen and its citizens
- Moscow is getting smarter and smarter
- Singapore has been ranked the world's smartest city. What does a smart city look like on the streets and in the homes of Singapore?
- In India, Bhopal is ranking #1 in the new ranking released by the Union Ministry of Housing and Urban Affairs (Times of India, 7 February 2021.)
- Dehradun smart city (India): Discover major flagship pan-city solutions
- Dholera smart city: One of the first smart cities in India

While data sharing is essential, opening the vault also expands the cyber-attack surface area.

So, how do we keep data private from the masses while sharing it among stakeholders?

Can smart cities be secured and trusted?

In digital cities, connected cameras, intelligent road systems, and public safety monitoring systems can provide an added layer of protection and emergency support to aid citizens when needed.

- But what about protecting smart cities themselves from vulnerabilities?
- How can we defend against hacking, cyber-attacks, and data theft?
- In cities where multiple participants share information, how do we trust that participants are who they say they are?
- And how do we know the data they report is true and accurate?

The answers lie in physical data vaults and strong authentication and ID management solutions. Smart cities can only work if we can trust them.

Four core security objectives for smart city solutions

All ecosystem partners - governments, enterprises, software providers, device manufacturers, energy providers, and network service providers - must do their part and integrate solutions that abide by four core security objectives:

- **Availability:** Without actionable, real-time, and reliable data access, the smart city can't thrive. How information is collected, distilled, and shared is critical, and security solutions must avoid adverse effects on availability.
- **Integrity:** Smart cities depend on reliable and accurate data. Measures must be taken to ensure that data is accurate and free from manipulation.
- **Confidentiality:** Some of the data collected, stored, and analyzed will include sensitive details about consumers themselves. Steps must be taken to prevent unauthorized disclosure of sensitive information.
- **Accountability:** Users of a system must be responsible for their actions. Their interactions with sensitive systems should be logged and associated with a specific user. These logs should be difficult to forge and have reliable integrity protection.

Strong authentication and ID management solutions must be integrated into the ecosystem to ensure that data is shared only with authorized parties to achieve these security core objectives.

- The solutions also protect backend systems from intrusion and hacking.
- Thankfully, legislation is being introduced to address threats and potential market failure due to growing digital security concerns.

Like the IoT Cybersecurity Improvement Act in the U.S. signed on 4 December 2020, or the UK IoT security law (not passed yet in June 2021), legislations will help establish minimum security requirements for connected devices.

- Show me the Money: how do we monetize smart cities?
- In the age of IoT and smart cities, data is the new oil.
- For smart cities to thrive, we need to establish **sustainable commerce models** that facilitate all ecosystem players' success.
- The software must be woven into the fabric of IoT solutions to benefit all ecosystem contributors; this includes OEMs, developers, integrators, governments, etc.
- Each member's intellectual property needs to be valued and rewarded.
- Subscription software capabilities enable new business models that allow each contributor to extract value from their contribution to the ecosystem.

Subscription-based models offer a way to monetize hardware and software to build smart infrastructures and spread out expenses moving away from a substantial one-time CAPEX spend.

- Expensive medical equipment like MRI scanners, for example, can be sold on a **cost-per-scan** basis rather than as a one-time upfront expense for hospitals. This creates a win-win situation for hospitals and suppliers alike.
- And one day soon, cities will offer affordable subscriptions to fleets of **vehicles shared** between owners who may choose from an array of custom options. This move could radically reduce traffic and optimize traffic patterns and ride-sharing.

As urban areas continue to expand and grow, **smart city technology** is developing alongside enhancing sustainability and better serve humanity.

By leveraging pervasive connectivity, open data, end-to-end security, and software monetization solutions, we can align evolving smart city needs for a much-improved experience for all ecosystem partners.

SMART CITY COMPONENTS

Many components fall under the umbrella of smart city infrastructure. When used singularly or in conjunction with multiple initiatives, improvements can significantly enhance operational efficiencies and the quality of life for residents. There is no one model for a smart city, but key aspects include the following:

Citizen Engagement

Citizen engagement is crucial to large-scale infrastructure projects. Local citizens need to be aware of and prepared for any changes. When citizens and municipal governments work together, positive community changes are possible.

Infrastructure

Improving the efficiency and capabilities of government services is key to preparing for and maintaining sustainable growth. Critical infrastructure services include city-wide facility management, wastewater treatment, and water distribution.

Broadband

The internet has become an essential service—similar to electricity. As a result, there is a critical need for information communication technology and public Wi-Fi. High-speed internet connections allow for economic development and the growing remote work trend.

Public Safety

In terms of technology, many tools can help sense, analyze, and act on potential public safety threats. For example, smart cameras can detect accidents and call for EMS and microphones can identify gunshots, triangulate a shooter's location and relay that information to the closest first responder.

Resiliency

Cities need to be resilient to any challenge they may face, particularly concerning critical infrastructure services that citizens rely on. If faced with rising water, inclement weather, or extreme conditions, technologies like renewable energy, smart grids, centralized building automation, communication redundancy, and asset efficiency can help cities avoid critical failures.

Energy Efficiency

Increasing energy efficiency isn't just good for the environment. It's also good for a city's bottom line. Connected buildings, LED lighting upgrades, smart street lighting, building envelope improvements, HVAC equipment upgrades, and energy dashboards can deliver financial benefits to public budgets through increased income and decreased operating expenses.

Healthier Communities

Enabling healthcare access for all and nurturing the social, economic, and environmental factors contributing to residents' overall well-being is a common goal for most cities. Telemedicine opportunities for homebound individuals are an excellent example of how digitized cities can better serve citizens.

Transportation

Enabling intelligent transportation systems or using traffic management centers is vital for cities' growth. For example, traffic management centers can monitor and coordinate large networks of sensors to improve traffic congestion. Additionally, with the expected surge in all-electric vehicle (EVs) ownership, cities will need to increase charging station availability to make charging easy, accessible, and equitable. Not only will this benefit citizens, but it will help cities reduce carbon emissions, meet CO2 reduction goals, and reduce harmful pollutants.

Data Analytics

Access to better data management and an understanding of the 3 Vs (volume, velocity, and variety) will enable cities to make better, faster decisions. With actionable data insights, policymakers can improve operations, reduce resource consumption, monitor security risks, and better manage budgets.

Budget-Neutral Financing

Public officials understand that every smart infrastructure improvement they want to make has a funding obstacle to overcome. What many smaller cities and towns don't know is that budget-neutral financing is possible in most states. With state-enabled legislation and federal aid from the American Rescue Plan Act (ARPA) and the Infrastructure Investment and Jobs Act (IIJA), many cities and towns can make these smart infrastructure improvements with little to no up-front capital required.

Bringing it All Together

With strategic planning and commitment to innovation, local governments undertaking improvement projects in these areas will undoubtedly take a step toward becoming smarter cities and towns. Doing so should yield positive outcomes, not just in terms of growth, but will help provide sustainable economic development opportunities, improved government efficiencies, healthier environments, better traffic flow, increased safety, and upgraded infrastructure.

COMMUNICATION INFRASTRUCTURE FOR THE SMART CITY

Fibre, 5G, WiFi, Bluetooth, Lora, Sigfox, etc. Which is the ideal network for connecting the smart city's many communicating objects? None of the above, or rather all of them, say the experts. Axians (VINCI Energies) Business Unit Manager Stéphan Guidarini says that "There is no single, universal solution for connecting all the buses, cars, urban lighting, and water and energy systems that have built-in sensors.

The best solution for a municipality is a combination of mobile, fibre, WiFi, and Lora. Each technology meets a specific need, in terms of coverage, reliability, and efficiency." The choice of technology will depend on the priority - range, cost, energy consumption, bandwidth, latency, or data transfer rate.

Analytical smart camera

When a fibre optic network has already been rolled out in a municipality, it can be used to serve different points if a high data rate is needed - for example, to install analytical cameras, which can sometimes be more advantageous than a network of sensors. In Lisbon, Axians has rolled out a smart parking solution based on the technology, which combines video image capture and analysis not just to report information on parking space availability but to detect inappropriate vehicle behaviour or wrong-way traffic.

Traffic analysis at the Place de la Nation in Paris is another example of the use of the smart camera, which can recognise pedestrians, cars, and even abnormal vehicle behaviour within an urban area to detect an accident. "The analytical camera is more precise and easier to install. It will come into widespread use in the smart city," says Stéphan Guidarini. This is particularly true now that edge computing can be used to process data locally and avoid reporting data that could be used to identify people or vehicles.

Promising 5G

Bandwidth is needed to transfer these large files and sometimes WiFi is not enough. If, on the other hand, the municipality has decided to offer its residents this wireless Internet connection service, the network could also be used to interact with the citizen installing the app. WiFi service could also be used to count pedestrian traffic flows, as is now being done in Aix-en-Provence, in order to study how people move within the city.

For services using sensors that transmit relatively little data, low data rate networks that consume little energy, such as Lora and Sigfox, are appropriate. In Aix-en-Provence, sensors measuring such things as air quality and waste-bin-filling rates send the data once an hour to the city's Lora network. The two-way network is also used to send instructions to connected objects to control such things as smart watering of city gardens. Lora is a technology based on open protocols, enabling the municipality to upgrade its smart city scenario over time while protecting its long-term investment. The use of open protocols is recommended since interoperability is key to ensuring smart city efficiency.

5G technology, which has a promising future in the smart city, will not replace the other networks. Its higher cost will mean that it will be used for purposes requiring high data rates and high performance – for example, to manage mobility involving the connected vehicle, says the Axians business unit manager.

In addition to technical features, it is important to consider other criteria when rolling out a city's communication infrastructure. First, "Start with the city's existing systems." Existing communication networks can be re-used in the new infrastructure. Then, to keep costs under control over time and ensure that data governance is independent of the stakeholders in the region, it is preferable for the city to own its communication networks..

Communication networks are the first of the city's three basic technological components. "The second is the pooled database that stores and correlates all city data reported by the networks. Data centralisation is crucial in a Smart City programme because it increases the value added delivered by the third component – the applications that support new uses for citizens and city departments. The better these applications combine data from multiple sources to present a broad picture representative of the way the city actually works, the more useful they will be," says Edouard Henry-Biabaud.

AN INNOVATIVE SCANNER FOR BUILDING INSPECTIONS

The sensors form a network which use electromagnetic waves which detect subsurface anomalies, across a range of materials including tiles, wood, concrete, PVC and dielectric composites. The data transmitted by these sensors are processed by AI algorithms in the cloud. This provides technicians with information regarding various materials and their structural integrity. Additionally, 3D images of equipment embedded in walls can be created using magnetic resonance imaging (MRI).

WaveScan was originally designed in Singapore, and has been developed over the past decade. Originally, the system was used in industries such as aviation, where users were able to identify submillimetre defects such as cracks and corrosion holes in aircrafts, as well as in other areas such as in oil and gas pipelines. These industries are able to identify and address potentially disastrous faults, preventing devastation. The technology was then introduced into the construction industry, for example, allowing inspection of the facades of skyscrapers in Singapore.

These 3D scanners and AI form a more efficient system, therefore streamlining building inspections and work on equipment, reducing overall cost and increasing accuracy.

FUTURE VISION GENERATION

Constructing the future vision entails defining and describing a desirable future in which the problems and issues identified in relation to existing sustainable urban forms have been solved by meeting the stated objectives and thus achieving the specified goals and targets described in step 1. In general, future vision construction is about identifying the desired future state, which consists of vibrant descriptions of audacious goals and targets, as well as reflective statements addressing the aspired future.

On the visionary approach

The future vision is a result of the concept of urban sustainability as clarified, advocated, and advanced by many scholars, academics, theorists, and practitioners in the field, and demonstrated in numerous real-world cities across the globe, especially within ecologically advanced nations. According to several rankings, Sweden, Norway, Finland, Germany, and the Netherlands have the highest level of sustainable development practices. However, the development of the novel model for smart sustainable cities of the future is supported by several case studies from Sweden as well as their integration in terms of the planning practices and development strategies through which sustainable urban forms can be achieved.

Additionally, this model involves the way instrumentation, datafication, and computerization are opening up dramatically different forms of optimizing and enhancing the performance of such forms, thereby increasing their contribution to the goals of sustainable development. This entails the ways in which the informational landscape of smart cities as underpinned by big data technologies and their novel applications can be integrated with the physical landscape of sustainable cities, and what this implies in regard to increasing their sustainability benefits. The essence of the idea revolves around the need to harness, analyze, and leverage the deluge of urban data that has hitherto been mostly associated with smart cities but has clear synergies in the functioning, planning, and development of sustainable cities in terms of improving, advancing, and maintaining their contribution to sustainability.

The problems and issues that the sustainable city faces today will, especially if its landscape and strategy continues to be extremely fragmented from and weakly connected with those of the smart city at the technical and policy levels, increase in the future with possibly much greater compounding effects due to the rapid urbanization of the world and the mounting challenges of sustainability in a rather increasingly technologized and computerized world. As a scholarly endeavor, the development of the novel model for smart sustainable cities of the future as a holistic approach to city planning and development is primarily aimed at bringing together and interlinking the sustainable city and smart city landscapes and strategies so as to address and overcome a set of challenging problems associated with the existing sustainable urban forms.

This requires finding more creative and effective ways of merging sustainability knowledge with advanced technologies to enhance the performance of such forms in the face of urbanization using cutting-edge technologies. This can be accomplished by amalgamating the compact city with the eco-city into one model of sustainable urban form in terms of the underlying typologies and design concepts as planning practices, and then augmenting this model with big data technologies and their novel applications as a set of innovative solutions and sophisticated approaches being offered by the data-driven city.

In this respect, city operating system, operations centers, innovation and living labs, and strategic planning and policy offices will handle the activity of generating, processing, and analyzing the data deluge aimed at adopting those innovation solutions and sophisticated approaches in the context of the smart sustainable city. Practical uses and applications in this regard span a range of urban systems and domains in terms of operations, functions, services, designs, strategies, and policies with respect to sustainability.

The future vision has a high expectation on big data technology to deliver the needed solutions and approaches to meet the optimal level of sustainability and enable the built environment to function in a more constructive way than at present in terms of lowering energy consumption, mitigating pollution, and minimizing waste, as well as in terms of improving equity, inclusion, and the quality of life. This is to be determined by whether and the extent to which a given city is currently badging or regenerating itself as, or manifestly planning to be, sustainable or smart sustainable. And what this entails in terms of long-term targets of sustainable development as set by that city, in particular in connection with its design concepts, typologies, spatial organizations, and scale stabilizations as planning practices. In the short term, although big data technology could theoretically help meet the optimal level of sustainability and enable the instrumentation, datafication, and computerization of the built environment towards purposeful urban functioning and planning, this would be difficult and expensive. Nevertheless, the future vision can be feasible because it has to be realized over the long run.

The technological vision is based on the assumption of a full development, integration, and deployment of big data computing and the underpinning technologies which exist today and are likely to become widely available in the years ahead to achieve the sought goals. The incorporation of these advanced technologies into urban environments is supported by their untapped potential for and proven role in overcoming the problems and challenges of urbanization and sustainability. In this respect, big data computing and the underpinning technologies will be determining in the process of redesigning and restructuring urban places to achieve the optimal level of sustainability.

The future vision

Taking the prevailing and emerging trends to the extreme with the main expected developments in mind, we singled out one major societal driver for one scenario: a situation that is most likely to happen in the future:

> A scenario where innovations and advancements in big data science and analytics and the underpinning technologies as a disruptive form of science and technology dramatically changes the rules by which society functions on a global scale.

Accordingly, the futures study envisions the smart sustainable city as:

A form for human settlements that will be able to improve, advance, and maintain its contribution to the goals of sustainable development by being pervaded, monitored, understood, and analyzed by advanced ICT. As such, it is to be realised by the planning practices and design strategies pertaining to the most advocated and prevalent models of sustainable urban form as integrated—as well as underpinned by big data computing and the underlying core enabling technologies in terms of the instrumentation, datafication, and analytics of the built environment.

Related sophisticated approaches and novel applications will be developed, applied, and enhanced by a number of strategic urban actors, including urban operations centers, urban services agencies, strategic planning and design offices, policymaking bodies, research centers, and innovation and living labs. The main strategic goal of the future model of data-driven smart sustainable urban form is to secure and uphold environmentally sound, socially beneficial, and economically viable development towards achieving sustainability.

In light of the above, envisaging the smart sustainable city of the future focuses on the urban and technological components and how they should be integrated that make the city functions as a smart sustainable entity as well as a social organism. Central to this quest is the idea of big data computing and the underpinning technologies as an advanced form of ICT penetrating wherever and whatever it can of the built environment to improve and sustain the performance of what and how urban stakeholders can envision and enact in terms of new forms of cities with regard to sustainability.

Furthermore, advanced ICT comes into play as a response to the commonly held view that cities should be conceived in terms of both urban strategies and processual outcomes of urbanization, which involves questions related to the behavior of inhabitants; the processes of living, consuming, and producing; and the processes of building urban environments—in terms of whether these are sustainable. The underlying assumption is that conceiving cities only in terms of, or accounting only for, urban strategies to make cities more sustainable remains inadequate to achieve the elusive goals of sustainable development.

As hinted at above, the novel model for smart sustainable cities of the future, the more detailed version of the future vision, integrates two models of sustainable urban form: the compact city and the eco-city, with the data-driven city. This will result in a holistic approach to urbanism, which is different, to a great extent, from these cities taken separately as existing approaches to urbanism. Worth pointing out, to reiterate, is that the focus of this amalgamation is on the design concepts and typologies characterizing both the compact city (i.e., compactness, density, diversity, mixed-land use, and sustainable transport) and the eco-city (i.e., renewable resources, passive solar design, ecological and cultural diversity, greening, environmental management, and other key environmentally sound policies) together with the innovative solutions and sophisticated approaches being offered by big data technologies and their novel applications for sustainability, which relate to the data-driven city and its components (i.e., urban operating centers, research centers, living labs, and innovation labs). The nature and scope of this amalgamation are to be determined by how and the extent to which the characteristic features of the data-driven city would dovetail with those of the integrated model of sustainable urban form towards producing what can be described as—data-driven smart sustainable urban form.

Furthermore, it must be noted that there are neither real examples of a truly smart sustainable city that have actually been delivered and thus no precedents to reference, nor future-proofing of the big data technology to ensure that it is able to be adapted, modified, and built upon in an effective way over the next 25 years or so in response to the dynamic changes of technology and fast-moving hi-tech industry. Therefore, the planned big data technology solutions must be evaluated through actual implementation and its successfulness in order to outline the actual opportunity pertaining to the improvement and advancement of urban sustainability. Indeed, big data computing and the underpinning technologies intended to support the smart sustainable city of the future are currently evolving along with those experts and professionals who are needed to support and operate them; sustainability objectives, goals, and directives are increasingly being, and should continue to be, supported and facilitated using this advanced technology as much as possible across urban domains in terms of operations, functions, services, designs, strategies, and policies; and citizens and communities must be involved and engaged with big data technology and related platforms on a far broader scale. The road ahead promises to be an exciting one as more cities become aware of the great potential and clear prospect of integrating the smart city and the sustainable city as landscapes and strategies. In the sequel, we describe the three strands that comprise the novel model for smart sustainable cities of the future as hinted at in the description of the vision of the future above.

SUSTAINABLE CITIES

There are multiple views on what a sustainable city should be or look like and thus various ways of defining or conceptualizing it. Generally, a sustainable city can be understood as a set of approaches into operationalizing sustainable development in, or practically applying the knowledge about sustainability and related technologies to the planning and design of, existing and new cities or districts. It represents an instance of sustainable urban development, a strategic approach to achieving the long-term goals of urban sustainability. Accordingly, it needs to balance between the environmental, social, and economic goals of sustainability as an integrated process. Specifically, as put succinctly, a sustainable city "strives to maximize the efficiency of energy and material use, create a zero-waste system, support renewable energy production and consumption, promote carbon-neutrality and reduce pollution, decrease transport needs and encourage walking and cycling, provide efficient and sustainable transport, preserve ecosystems and green space, emphasize design scalability and spatial proximity, and promote livability and community-oriented human environments."

There are different instances of sustainable cities as an umbrella term, which are identified as models of sustainable urban forms, including compact cities, eco-cities, sustainable urbanism, green urbanism, new urbanism, and urban containment, with the first two being often advocated as the most sustainable and environmentally sound models. In addition, Jabareen ranks compact cities as more sustainable than eco-cities from a conceptual perspective using a thematic analysis. However, the effects of these models are compatible with the goals of sustainable development in terms of transport provision, mobility and accessibility, travel behavior, energy conservation, pollution and waste reduction, economic viability, life quality, and social equity. Furthermore, there are multiple definitions of compact cities and eco-cities in the literature. These definitions tend to be based on the wider socio-cultural context in which these models of sustainable urban form are embedded in the form of projects and initiatives and related objectives, requirements, resources, and capabilities.

CHAPTER 10

TECHNOLOGIES: THE FINE PRINT

Most of current urbanism is based on a technological paradigm in which the quantification of elements such as housing, jobs and parking spaces, standards, and regulations seem more important than achieving resilience. These kinds of urbanism are strongly economically driven and money-based. Recent developments, such as smart cities, with their focus on data and technology, often deepen this technological paradigm, hence adding vulnerability to urban systems. The urban social and natural systems are seldom capable of dealing with sudden shocks, which are bound to occur at an increasing rate.

In the future we need to deal with a triplet of uncertainties:

- The first of this triplet are developments, which are uncertain by nature. Examples of these, often referred to as wicked problems, are climate change, mass migration, and disruptive technological developments. These are impossible to predict and are therefore inherently uncertain;
- A second element consists of enforced uncertainties. Several sustainability transitions, such as the transition towards a green economy or a low-carbon energy system are deliberately enforced by international agreements and national policies. This type of uncertainties will impact the urban system. In order to accommodate these transitions the city itself needs to transform;
- Thirdly, there is uncertainty that stems from being exposed to the aforementioned two elements: exposure to uncertainties. For instance, the urbanisation of the global population continues, which means an increasing number of people live in cities. At the same time, most cities worldwide are located in vulnerable, exposed areas. Together, this means that the number of people that will have to deal with uncertainties is rapidly increasing.

Because of this triplet of uncertainties, a definition of the future of eco-urbanism is required: urbanism that deals with the above-mentioned triplet of uncertainties, using both advanced technologies, the laws of nature, and self-organizing adaptive systems.

THE HISTORY OF SUSTAINABILITY IN URBANISM

Looking back in time, sustainability has not always been a central theme in urbanism. 'The term 'sustainable urbanism' encompasses topics of sustainability related to the entire process of city development and management, while sustainable urban design sits somewhere within that. Sustainable urban design is not necessarily a clearly delineated subset of sustainable urbanism, but instead can be thought of as focus area within it that is concentrated around issues of design, while still maintaining strong links to the other realms such as planning, engineering, real estate, and policy.

The famous image of the architect eating the landscape, by Malcolm Wells, illustrates the way urbanism was seen for a long time: as a destroyer of natural values. Girardet was one of the first to acknowledge the importance of an integrated approach to developing cities in a sustainable way. He connected themes such as looking at cities as ecosystems, the footprint of cities, urban heat islands, and others with urban design schools, such as the garden city movement, the modern city of Le Corbusier, Frank Lloyd Wright's Broadacre city, and megacities in general. His pledge for convivial cities, the role of citizens and the local scale, and the attention for health, greening the city, renewable energy and recycling, and sustainable forms of mobility, are still relevant.

Since then, many perspectives, visions, and conceptual practices have been developed, each of them taking sustainability into account in the planning and the design of cities. In the last 20 years, a number of theories or ideologies have influenced thinking about (sustainable) urbanism. Scholars such as Farr, acknowledging a strong shift towards integration of ecological analyses in urbanism, Beatley, focusing on ecologically-based land use planning and integrating nature into urban design and planning, Mostafavi, deliberating on ecological urbanism, Waldheim, launching landscape urbanism in which landscape supplants architecture as the basis of urban design, and Newman, elaborating on the resilience of the city, as well as Duany, proposing an extensive manual on smart growth, and Haas, defining a wide range of principles for planning of sustainable and resilient cities, have all covered ground related to building a framework for sustainable urbanism.

However, much of the literature on sustainable urban design still lacks grounds for holistically relating the different aspects of urban design. Taking these readings as the starting point of thinking, sustainability in urbanism has been divided in seven distinctive periods which illustrate how the concept of sustainable urbanism has evolved over time.

From the early days when aesthetics were defined by Vitruvius as not only beautiful (venustas), but also solid (firmitas) and useful (utilitas), urban sustainability is defined as a multi-layered concept and aesthetics is equally important as economic viability or ecological value, not more or less important. Sustainability is not similar to environmental quality. A high environmental quality alone does not guarantee a high degree of sustainability. Similarly, an urban environment with a high environmental quality or a high economic value, but without beauty is not considered sustainable either.

After this, periods such as rationalism and conceptualism can be distinguished. Rationalism assumes knowledge will implicitly lead to solutions. An expert planner or the designer 'who knows it all' is required to translate this knowledge into functional urban environments. The shape of the city follows the land use, or 'form follows function', and engineers play an important role in the final outcome of the design process. In this context sustainability becomes a quantifiable aspect and standards and regulations were introduced, such as for clean water, air, and soil. In the conceptualism period, the water system, the soil, and nature are used as layers, an approach that is widely used in the Netherlands.

The abiotic, biotic, and anthropogenic layers shape the landscape. The layers aim to separate different dynamics of use. Where the subsoil is seen as changing extremely slowly, and the networks change at a moderate pace, the occupation patterns can change relatively rapidly. It assumes the occupation patterns are nested in the infrastructure networks, which in turn are embedded in the substrate system. In the casco-concept, which is based on this thinking, these spatial dynamics are separated. In the concept called 'the strategy of the two networks', higher dynamic uses (traffic, industries, intensive forms of agriculture) are linked to the transportation network and the lower dynamic uses (nature, green, water, residential) to the water network.

In the negotiatism period, sustainability is seen as a discussable ambition in the planning process. New methods are developed to target sustainability ambitions, such as the DCBA-method, in which every urban design aspect requirement is set for level D to A, D being the highest. Over time, the negotiation process evolved into advanced participatory processes, in which engaged citizens could co-design and co-create the sustainability of their direct environment. In these sessions, the wisdom of the crowd is used to identify future strategies for a sustainable spatial development.

Looking at the city as an ecosystem is the key of the period in which urban metabolism defines the city in terms of its flows. When the layer approach is combined with this system's thinking, a comprehensive model for urban environmental design emerges. The degree to which the system is able to deal with the flows, or networks of traffic, water, energy, and materials, determines the level of sustainability of the system. The design of the city is based on ways in which depletion of resources and production of waste are reduced, and in which resources can be reused and recycled within the city. The most recent periods, emergism and anti-fragilism focus on the city as a responsive system, in which self-organisation and the adaptive capacities of complex systems determine urban processes. Emergism takes complexity as the input for the design of cities. Self-organisation and emergence are key concepts, and are used to design interventions in the system to achieve certain changes. These concepts are common in nature and can be used in designing future cities and landscapes that are more adaptable.

The designer acts as a facilitator of the process of change, intervening at specific places or times to initiate a change in the system. Approaches such as eco-acupuncture and Swarm Planning aim to design small interventions in an existing urban precinct or landscape to transform the area and become more resilient. In order to apply these principles, physical elements in the city or the landscape need to be made moveable and capable of responding to different paces of change that might occur in the city: fast, slow, or sudden. In order to adapt to new circumstances urban systems need to become agile.

Rather than only responding to change by coping with it, urban environments can actually become stronger than before through their response to climate events. This concept is called anti-fragility. 'Ant-fragility is defined as a convex response to a stressor or source of harm (for some range of variation), leading to a positive sensitivity to increase in volatility (or variability, stress, dispersion of outcomes, or uncertainty, what is grouped under the designation "disorder cluster")', and offers interesting opportunities. For instance, if the city can create a stronger flood defence by itself once a flood occurs, the city becomes safer. Anti-fragilism is not a common approach in urbanism yet. However, the current combination of increasing uncertainties and unalterable cityscapes places urban population increasingly at risk. Therefore it is necessary to investigate the solutions anti-fragility might bring to urbanism.

Learning from the history of sustainability in urbanism, several key-concepts stand out. The concepts developed to create sustainable framework with the landscape as a basis, such as the layer approach, the casco-concept], and the Two Networks strategy, have determined much of the work and research being carried out in landscape architecture.

The engagement of communities in the planning and design of urban environment is another key concept which determines the sustainability, support, and acceptance of urban change amongst residents. Recent concepts such as co-design and co-creation allow for a broad spectrum of expertise sets in the design process, which stimulates the development of new and innovative design ideas and new knowledge.

The view of the city as an ecosystem has changed the way we look at urban change and design. The metabolism and design of flows made it possible to discuss the sustainability of the city as a whole, and close the cycles of energy, water, and materials.

More recently, the city is seen as a complex adaptive system and this implies that adaptation and resilience of the city can be discussed as core characteristics. The adaptivity of the city can be influenced through strategic design interventions supporting self-organisation. Adaptation requires creating space to adapt, hence this concept advocates redundancy ('space for the unknown') in the urban realm.

A future avenue to increase the strength of the city is to create anti-fragile environments, which grow under influence of external impacts.

They lead to six principles that serve as a basis for design: closing cycles, innovative designs, create space for the unknown, anti-fragility, people's engagement, and taking the landscape as basis for design. These principles are regarded under the larger comprehensive concepts such as society, complexity, and landscape.

ECO-INITIATIVES IN THE BUILT ENVIRONMENT

Drivers for Green, or 'Ecological', Urbanism

Green urbanism is, by definition, interdisciplinary, as in 'ecological design', which has been and will be interdisciplinary, not the property of any single profession or movement. Where ecologists look at the size of the ecological footprint, public health focuses on the exposure to pollution, and economists have their eye on the differences in price of real estate, green urbanism unites these views. Green urbanism makes the effort to minimize the use of energy, water, and materials at each stage of the city's or district's life cycle. In a new era of uncertainty energy, water and food supply are critical.

Humans create a world of increasingly non-calculable uncertainty with the same speed of its technological developments. For eco-urbanism, six driving forces are distinguished: They can be driven by environmental challenges, emerge out of socio-economic pressure, are seen as a business opportunity, be part of cultural branding, evolve out of political leadership, and could be instigated by international co-operation. The development of the eco-city, based on a different combination of these drivers, emerged in three phases (Table 1): after a normative and regulatory phase, the current discourse is based on innovation.

Table 1. Three phases of eco-city development

	Phase	Perspective	Means
1980's	Grassroots, visions	Normative	Environmental standards
1992–early 2000	Local and national	Regulatory, standardised	Standardized models
	experimentation		
2000–present	Global expansion, mainstreaming	Innovation, conceptual	Technology and design

These three phases can be linked, with differences in the means to realize an eco-city: Environmental sustainability is used in the normative phase, models are used to standardize in the second phase, and, in the current period, sustainability is achieved through technology and design.

Urban Design Principles

Based on these driving forces for green urbanism, a series of principles can be defined (Table 2) and applied in a systemic and adapted way, requiring a specific approach for every unique site. The future of our society is not merely a technical matter of finding more eco-friendly energy solutions, but a question of holistic environmental and social sustainability, and identifying principles for healthy communities. Much of Green Urbanism is common sense urbanism.

Table 2. Principles for Green Urbanism

Principle	Aspects
Climate and context	Site's climate conditions, orientation and compactness, landscape, topography and resources, maintain complexity, own methods, and tailored strategies for every district take advantage of local potentials
Renewable energy	Availability of local renewables (energy potentials [72], city district as a producer of energy, distributed energy supply though decentralized systems, local storage, smart grid, energy efficiency, co-gen, cascading exergy principles, intelligent building management.
Zero-waste	Circular, closed loop, turn waste into resource, reduce, reuse, recycle, and compost waste to produce energy, remanufacturing, balance in nitrogen cycle
Water	Reduce consumption, efficiency of use, water quality, aquatic habitats, city as a water catchment area, storm water retention and flood management, rainwater harvesting, local treatment of waste-water, safe water and sanitation, algae, integrated urban water cycle planning, black and grey water treatment, dual water systems, drought resistant crops
Landscape and urban biodiversity	Local biodiversity, habitat, ecology, wildlife rehabilitation, forest conservation, urban vegetation, inner city gardens and urban agriculture, mitigating Urban Heat island effects, tree planting, restoring stream and river banks, de-pavements, carbon storage
Sustainable transport and good	Public space network, post fossil mobility, access to public
public space in compact poly-	transport, safe bicycle ways, smart vehicles, walkeable city,
centric cities	streetscapes for healthy active lifestyle, medium dense housing typologies

Local materials with less	Advanced materials technology, local materials, modular
embodied energy	prefabrication, lightweight structures, disassembly, resource recovery, reuse building components
Density and retrofitting existing	Densification of the city, mixed use urban infill, retrofitting
districts	inefficient building stock, better land use planning, public space upgrading, adaptive reuse, city above the city, self-sufficiency
Passive design for buildings	Low energy, zero-emission design, reduce energy use, compact
and districts	solar architecture, bioclimatic architecture, design for disassembly, solar architecture, flexibility in plans, energy generating buildings
Liveability and mixed use	Affordable housing, healthy community, social inclusion, mixed use, liveable and flexible housing typologies, secure tenure, diversity, integrating a diversity of economic and cultural activities,
Local food & short supply chains	Local food production, regional supply, urban farming and agriculture, community and allotment gardens, roof gardens, urban market garden, paper bags, recycling, organic produce
Identity and sense of place	Public health, cultural identity, urban heritage, air quality, distinct environment, grassroots strategies, affordable studio space, creativity of government and citizens, health, activities and safety
Urban governance and leadership	Evolutionary and adaptive policies, decision making, and responsibility shared with empowered citizenry, enabling citizens, updating building codes, improve planning participation, integrated public awareness, legislating controls on density and urban sprawl, support high quality densification, finance low-to-no-carbon pathways, eliminate fossil fuel subsidies, certify urban development projects
Education, research & knowledge	Up-skilling, knowledge dissemination, primary and secondary school teaching programs, university as a think tank, redefine education of architects, foundation of centre for sustainable urban development

Eco-City Typologies

'Full awareness of problems posed by climate change and consumption of fossil fuel is of equal importance and just as far reaching, raising the possibility of entirely new city models and typologies that are likely to emerge'. For instance, Kahn defines a green city as a metropolis, which has a core area containing a substantial population nucleus, surrounded by adjacent communities with a high degree of social and economic integration.

Green stands for cities that have clean air and water, pleasant streets and parks, are resilient in the face of natural disasters, and have a low risk at infectious disease outbreaks, and they encourage green behavior.

Haughton distinguishes the following four models of cities and their relations to their environmental hinterlands:

- **Self-reliant cities:** intensive internalization of economic en environmental activities, circular metabolism, bioregionalism, and urban autarky;
- **Redesigning cities and their regions:** planning for compact, energy efficient city regions;
- **Externally dependent cities:** excessive externalization of environmental cost, open systems, linear metabolism, and buying in additional carrying capacity;
- **Fair Share cities:** balancing needs and rights equitably with regulated flows of environmental value and compensatory systems;
- Focusing more specifically on sustainable urbanism, Dc Jong distinguish an additional six types of cities:
- The sustainable city, which has a self-sufficient economic, social, and environmental system;
- The low carbon city is minimizing the human-inflicted carbon footprint by reducing or even eliminating the use of non-renewable energy resources;
- The smart city, in which investments in human and social capital is coupled with investment in traditional (transport) and modern information and telecommunication infrastructure, generating sustainable economic development and a high quality of life while promoting prudent management of natural resources;
- The knowledge city integrates cities that physically and institutionally combine the function of a science park with civic and residential functions;
- The resilient city in which a predictive model of urban systems is built through three interconnected components of uncertainty: adaptation, spatial planning, and sustainable urban form (compactness, density, mixed land use, diversity, passive solar, greening, and renewal and utilization), in order to deal with ecological problems, hazards and disasters, and shocks in urban regional economies, through promoting urban governance and institutions;
- The eco-city aims to reconstruct cities in balance with nature, a city built within the principles of living within the means of the environment.

An eco-city is a human settlement modelled on the self-sustaining resilient structure and function of natural ecosystems. The eco-city provides healthy abundance to its inhabitants without consuming more (renewable) resources than it produces, without producing more waste than it can assimilate, and without being toxic to itself or neighbouring ecosystems.

Its habitants' ecological impact reflects planetary supportive lifestyles; its social order reflects fundamental principles of fairness, justice, and reasonable equity. Best practice examples of eco-cities in which these principles to a lesser or larger extend have been realised include the EVA Lanxmeer project in the Netherlands, Masdar ecocity in UAE, Hammarby-Sjöstadt in Stockholm, Sweden, and Hafencity in Hamburg, Germany.

The following dimensions of eco-cities are distinguished:

- **Sustainable development**: meet the needs of the present without compromising the ability of future generations to meet their own needs;
- Bioregionalism: a region governed by nature not legislature;
- Community economic development: communities initiate their own solutions to their common economic problems;
- Appropriate technology: enhance self-reliance of people;
- Social ecology: committed citizenry cutting across class and economic barriers to address dangers;
- Green movement: community self-reliance, quality of life, harmony with nature, and decentralization and diversity;
- Ubiquitous eco-city: international knowledge transfer, carbon discourse, smart city and ecological cities as economic cities.

The regenerative city relies primarily on local and regional food supplies, is powered, heated, cooled, and driven by renewable energy, reuses resources, and restores degraded ecosystems. The aim of the regenerative city is to be able to regenerate the natural resources it absorbs, consisting of four regenerations: Regeneration of Resources (From linear to circular flows), Regeneration of Natural Capital and Urban Ecosystems (from consuming to 'prosuming'), Regeneration of Urban Spaces (from Sprawled to Dense) and Regeneration of Communities (from Passive to Active engagement). ReGen villages engineer and facilitate the development of integrated and resilient neighborhoods that power and feed self-reliant families. The Internet of Things (IoT) enables thriving communities with surplus energy, water, and organic food in the aggregate become asset classes that can amortize and reduce mortgage payments.

Urban Design Characteristics of Eco-Cities

More detailed characteristics of sustainable urban environments can be distinguished, such as BedZED (Table 3), and as described by Kenworthy:

Table 3. BedZED's One Planet Living principles for design

Principle	Design
Health and happiness	Encouraging active, social, meaningful lives to promote health and wellbeing
Equity and local economy	Create safe equitable places to live and work, which support local prosperity and international fair trade
Culture and community	Nurturing local identity and heritage empowering communities and promoting a culture of sustainable living
Land and nature	Protecting and restoring land for the benefit of people and wildlife
Sustainable water	Using water efficiently, protecting local water resources and reducing flooding and drought
Local and sustainable food	Promoting sustainable humane farming and healthy diets high in local seasonal organic food and vegetable protein
Travel and transport	Reducing the need to travel, encouraging walking, cycling and low-carbon transport
Materials and products	Using materials from sustainable sources and promoting products, which help people reduce consumption
Zero waste	Reducing consumption re-using and recycling to achieve zero-waste and zero pollution
Zero carbon energy	Making buildings and manufacturing energy efficient and supplying all energy with renewables

- Compact mixed use urban form;
- Natural environment permeates the city's spaces;
- Transit, walking and cycling infrastructure;
- Environmental technologies for water, energy, and waste;
- City centers are human centers, with access other than automobile;
- High quality public realm;
- Legible, permeable, robust varied, rich, visually appropriate, and personalized physical structure;
- Innovation, creativity, and uniqueness create economic performance;
- Visionary planning through 'debate and decide'.

Learning from the review of eco-city literature, many different concepts, models, and principles arise. Depending on the phase of development, several urban typologies are distinguished, which each focus on different objects. The objects can be grouped in flows, climate and landscape, transport and densities, and social aspects. These groups can be translated in principles for urbanism, to close the cycles, base the design on the landscape, involve people in a co-creation process, and enhance innovation.

SMART (AND) SUSTAINABLE URBANISM: KEY CONTRIBUTIONS, ISSUES, AND CHALLENGES

When discussing sustainability, ICT, and urbanization and thus sustainable and smart practices and solutions for cities characterized by the rapid urbanization, reference is often made to three concepts: sustainable cities, smart cities, and smart sustainable cities. During the late 1980s the discourse on sustainable cities focused on the environmental, social, and economic sustainability of cities, and during the 1990s, the focus turned to the discourse on smart cities emphasizing ICT as a lens to see the future urban development. ICT has since gained recognition that it can contribute to transforming cities into spaces that can adapt to environmental, societal, and economic shocks. Compared to smart cities and sustainable cities, which have been around for quite sometime, smart sustainable cities have come to light over the past few years. Not until very recently, the smart sustainable urban development approach has attracted significant attention among contemporary urban scholars, planners, and policymakers. Its insertion, functioning, and evolution as a discourse, social practice, and techno-urban innovation are increasingly shaped and influenced by technological, societal, and economic factors: ICT industry consortia, research institutes, governments, policy networks, triple helix of university-industry-government relations, and recent quadruple university-industry- government-citizen relations, not least in ecologically and technological advanced societies.

While there is a growing interest in this flourishing, interdisciplinary field of research, the academic discourse of smart sustainable urban development within the relevant literature is still scant-yet rapidly burgeoning. Indeed, a few studies exploring the subject of smart sustainable cities have been published in mainstream journals. The case is evidently different from smart cities and sustainable cities as urban development strategies, which have witnessed a proliferation of academic publications and thus varied emphases of research and a large body of practices.

However, the speed at which the field of smart sustainable cities is gaining momentum and attracting attention gives a clear indication of its developmental path, blossoming nature, and future direction. In fact, this field of research comes as a natural pursuit within urban planning and development in light of the unsolved and unsettled issues and intellectual challenges pertaining to existing models of sustainable urban form in terms of their contribution to the goals of sustainable development, as well as the deficiencies and misunderstandings associated with existing smart city approaches in terms of their explicit incorporation of the goals of sustainable development.

Sustainable Cities (Urban Forms)

Sustainable development has significantly shaped and influenced the development of city models in terms of different dimensions of sustainability. Unquestionably, it has inspired and motivated a generation of urban scholars and practitioners into a quest for the immense opportunities created by the development of sustainable urban forms-i.e., the contribution that such forms can make as to lowering energy use and lessening pollution and waste levels, while improving human life quality and well-being.

Scholars from different disciplines and practitioners from different professional fields have, over the past two decades or so, sought a variety of sustainable city models that can contribute to sustainability and its continuous improvement. Compact city, eco-city, green urbanism, and new urbanism are the most widely applied sustainable city models. The underlying challenge continues to motivate and induce scholars and practitioners as well as policymakers and decision makers to work collaboratively to put forward new models for redesigning and rearranging urban areas across many spatial scales to achieve the required level of sustainability, especially in relation to integrating its physical, environmental, economic, social, and cultural dimensions.

The ultimate goal revolves around developing more convincing and robust models. This has been one of the most significant intellectual challenges and research endeavors for more than two decades. This implies that it has been difficult to, in addition to translating sustainability into the built form of cities, evaluate the extent to which the so-called sustainable urban forms contribute to the goals of sustainable development and to strategically improve this contribution. Indeed, existing models of sustainable urban form still pose several conundrums and raise numerous issues-when it comes to their development and implementation as to their contribution to the fundamental goals of sustainable development.

This pertains to limitations, uncertainties, paradoxes, and fallacies. The main premise is that urban systems have been in themselves complex in terms of their operation, management, assessment, and planning, so too are urban domains in terms of their coordination, integration, and coupling, all in the context of sustainability. To put it differently, whether in discourse, theory, or practice, the issue of sustainable urban form has been problematic and difficult to deal with, and research results tend to be uncertain, weak, limited, divergent, and not conclusive, particularly when it comes to the actual effects of the claimed benefits of sustainabilit.

Regardless, Neuman contends that conceiving cities in terms of forms remains inadequate to achieve the goals of sustainable development; or rather, accounting only for urban form strategies to make cities more sustainable is counterproductive. Instead, conceiving cities in terms of "processual outcomes of urbanization" holds great potential for attaining the goals of sustainable development, as this involves asking the right question of "whether the processes of building cities and the processes of living, consuming, and producing in cities are sustainable," which raises the level of, and may even change, the game. This process-driven perspective paves the way for a dynamic conception of urban planning that reverses the focus on urban forms governed by static planning tools; this holds more promise in attaining the elusive goals of sustainable development.

This argument provides valid reasons (supported with the above discussion) to believe that the conclusion it supports is true. This validity is in accordance with the fact that it is timely and necessary to develop and apply innovative solutions to deal with the challenges of sustainability and urbanization and how they affect each other in the context of sustainable urban forms. This would necessitate a blend of sciences for creating advanced engineering solutions, which ICT is extremely well placed to initiate given that it is grounded in the application of data science, computer science, and complexity sciences to urban systems and domains and related intractable problems. One way forward to smarten up sustainable urban forms and thus tackle the current complex issues in an increasingly urbanized word is to adopt the cutting-edge solutions being offered by ICT of the new wave of computing and the associated big data and context-aware technologies and their novel applications.

These solutions involve constellations of instruments encompassing sensing technologies, data processing platforms, cloud computing and middleware infrastructure, and wireless communication networks as well as their use within diverse urban systems and domains of sustainable urban forms in terms of collecting, storing, coordinating, managing, processing, and analyzing continuous data on urban activities, processes, and citizens as to the flow of decisions about the physical, infrastructural, spatiotemporal, operational, functional, and socio-economic aspects of such forms.

The whole idea is that more innovative solutions are needed to overcome the kind of wicked problems associated with sustainable urban forms. In this regard, the main and relevant question is how such forms should be monitored, understood, analyzed, evaluated, and planned to strategically advance their contribution to sustainability. In fact, with their domains becoming subtly interconnected and their processes highly dynamic, sustainable urban forms are relying more and more on sophisticated technologies, and in the near future, the core enabling technologies of ICT of the new wave of computing will be the dominant mode of monitoring, understanding, analyzing, evaluating, and planning many sustainable cities.

In addition, cities are increasingly seen as adaptive and data-centric systems, characterized by dynamic changes, complex interactions, and multidimensional effects, in contrast to what was the case for 50 years ago when cities were still very much seen as unchanging structures and relatively closed systems. Accordingly, context information and big data from various urban domains as a strategic asset constitute the fundamental ingredients for the next wave of urban analytics and planning.

In all, it has become highly relevant and important in an increasingly computerized and urbanized world to amalgamate the design and planning concepts and principles of sustainable urban forms with smart solutions and sophisticated methods for the purpose of strategically assessing and optimizing the contribution of such forms to sustainability under what is labeled "smart sustainable cities of the future." Worth noting is that for many contemporary urban scholars, theorists, and planners, the design concepts and typologies of such forms are necessary to be adopted and implemented to achieve the goals of sustainable development-irrespective of how intelligently the underlying urban systems and domains can function and be managed, designed, and planned.

TECHNOLOGIES, TECHNIQUES, PROCESSES, AND SYSTEMS

The contemporary city is evolving into becoming computerized on a hard-to-imagine scale due to the rapid development of ICT and its great potential to enhance urban operational functioning and planning. This is increasingly fueled by new discoveries in computer science and data science, coupled with the quick-paced ubiquity and massive use of computational and data analytics within a variety of urban domains to address the complex challenges of sustainability and urbanization facing major cities of the world.

This is manifested in the ongoing large-scale design, development, deployment, and implementation of sensor technologies, data processing platforms, cloud computing infrastructures, middleware architectures, and wireless communication networks across urban environments. In parallel, the increasing convergence, prevalence, and advance of urban ICT are giving rise to new faces of cities that are quite different from what has been experienced hitherto on many scales. This is increasingly boosted by data acquisition and storage, information processing, data networking, and intelligence decision support increasingly infiltrating urban systems as operating and organizing processes of urban life.

Accordingly, it has been suggested that the potential of monitoring, understanding, and analyzing the city through advanced ICT can well be leveraged in advancing its contribution to the goals of sustainable development. Indeed, the cities that are engaging in the new transition in ICT are getting smarter in how to become more sustainable. Besides, cities as complex systems, with their domains becoming more interconnected and their processes being highly dynamic, rely more and more on sophisticated technologies to realize their potential for responding to the challenge of sustainability and urbanization. Among these technologies are big data analytics and context-aware computing, which are rapidly gaining momentum and generating worldwide attention in the realm of smart sustainable urban development.

Big data and context information constitute the fundamental ingredients for the next wave of urban functioning and planning, especially in relation to sustainability. There indeed is a variety of potential uses of big data analytics and context-aware computing to address urban sustainability issues from the source thanks to the deep insights, intelligent decision-making processes, and efficient services delivery enabled by data mining, machine learning, and statistics and related modeling, simulation, and prediction methods. This points to new opportunities and alternative ways to develop, operate, and plan future cities.

The prospect of smart sustainable cities is becoming the new reality with the recent advances in and integration of ICT of various forms of pervasive computing and the underlying cutting-edge enabling technologies. Smart sustainable cities typically rely on the fulfillment of the prevalent ICT visions of the new wave of computing, where everyday objects communicate with each other and collaborate across heterogeneous and distributed computing environments to provide information and services to urbanites and diverse urban entities. The most prevalent forms of pervasive computing in relation to the urban domain are UbiComp, AmI, the IoT, and SenComp. Context-aware behavior and big data capability are considered as prerequisites for realizing the novel applications pertaining to such technologies. In all, the expansion of these computing trends as to the underlying technologies and applications are increasingly stimulating smart sustainable city initiatives and projects in ecologically and technologically advanced nations.

Virtually every urban aspect, process, activity, and domain is now open to data collection and processing and often even instrumented for data collection and processing: operations, functions, and services in terms of management, control, optimization, enhancement, planning, and so on. At the same time, information is now widely available on external states and events such as urban trends, environmental dynamics, socioeconomic patterns, and so on.

This broad availability of data has led to increasing interest in methods and techniques for inferring context knowledge as well as extracting useful knowledge from various forms and sources of data-the realm of context-aware computing and data science-for knowledgeable and strategic decision-making purposes. In all, data are being produced and warehoused, the computing power is available and affordable, the environmental pressures and socioeconomic concerns are alarming, and urbanization challenges are enormous.

The need to understand what constitutes the informational landscape of smart sustainable cities in terms of big data analytics and context-aware computing technologies presents an important topic and new direction of research in the field of smart sustainable cities of the future. The prominence lies in identifying the core enabling technologies and related key techniques and processes required to design, develop, deploy, and implement big data and context-aware applications for advancing urban sustainability.

Topical studies tend to only pass reference to such technologies or to largely focus on one particular technology as part of big data and context-aware ecosystems in the realm of smart cities. Moreover, empirical research on the topic, with some exceptions, is generally limited to case studies without the use of any common conceptual frameworks. In addition, relatively little attention has been given to the integration of big data analytics and context-aware computing as advanced forms of ICT in the context of smart sustainable cities.

This topic is a significant research area that merits attention, and this endeavor is a first attempt to address these two major strands of ICT of the new wave of computing in relation to the informational landscape of smart sustainable cities. This is to highlight that computers have become far more powerful, networks have become ubiquitous, and techniques and algorithms have been developed that can combine a large number and variety of sensors and connect various datasets to enable broader and deeper computational and analytical solutions than previously possible. The convergence of these phenomena is increasingly enabling many applications of smart computing and data science principles and big data analytics techniques.

The original contribution we make with this chapter is to review and synthesize the relevant literature with the objective of identifying and distilling the core enabling technologies of big data analytics and context-aware computing as ecosystems in relevance to smart sustainable cities, as well as to illustrate the key computational and analytical techniques and processes associated with the functioning of such ecosystems. In doing so, we develop, elucidate, and evaluate the most relevant frameworks pertaining to big data analytics and context-aware computing in the context of smart sustainable cities, bringing together research directed at a more conceptual, analytical, and overarching level to stimulate new ways of investigating their role in advancing urban sustainability. The proposed frameworks, which can be replicated and tested in empirical research, will add additional depth and rigor to studies in the field. We argue that big data analytics and context-aware computing are prerequisite technologies for the functioning of smart sustainable cities of the future, as their effects reinforce one another as to their efforts for bringing a whole new dimension to the operating and organizing processes of urban life in terms of employing a wide variety of big data and context-aware applications for advancing sustainability. The main motivation for this endeavor is to provide the necessary material to inform relevant research communities of the state-of-the-art research and the latest development in the field of smart sustainable cities in terms of the major technological components of their informational landscape, as well as a valuable reference for researchers and practitioners who are seeking to contribute to, or working toward, the design, development, and implementation of smart sustainable city applications. Especially, with vast amounts of urban data being now available, diverse entities in connection with every urban domain are focused on exploiting data for sustainable advantage.

THE CORE ENABLING TECHNOLOGIES OF BIG DATA ANALYTICS AND CONTEXT-AWARE COMPUTING

Like other application areas to which big data analytics and context-aware computing as advanced strands of ICT of the new wave of computing are applied, smart sustainable cities require these two related digital ecosystems and their components to be put in place, spanning different spatial scales in the form of enabling technologies necessary for designing, developing, deploying, and implementing the diverse applications that support, and ideally integrate, the dimensions of urban sustainability. As scientific and technological areas, these two strands involve low-level data collection, intermediate-level information processing, and high-level application action and service delivery. Worth noting is that as a result of the ongoing effort to realize and deploy smart sustainable cities, which are evolving due to the advance and prevalence of the enabling technologies of ICT of the new wave of computing, all the three areas are under vigorous investigation in the creation of urban environments merging the informational and physical landscapes of such cities for advancing sustainability.

There are many permutations of the core enabling technologies underlying big data analytics and context-aware computing. However, they all pertain to ICT of the new wave of computing, an integration of UbiComp, AmI, the IoT, and SenComp, which will in the near future be the dominant mode of monitoring, understanding, analyzing, and planning smart sustainable cities to improve sustainability. It is worth iterating that both big data analytics and context-aware computing share the same core enabling technologies because they are an integral part of ICT of the new wave of computing, as we will elucidate below.

As such, they involve unobtrusive and ubiquitous sensing technologies and networks, sophisticated data management and analysis approaches, data processing platforms, cloud computing and middleware infrastructures, and advanced wireless communication technologies. These are to provide solutions in the form of useful and context knowledge for the purpose of achieving the required level of sustainability in the context of smart sustainable cities.

Moreover, to have effective and successful solutions on the basis of core enabling technologies, it is required to select a number of design and development priorities in a planned manner prior to any deployment and implementation. For example, it is essential to consider flexible design, quick deployment, extensible implementation, more comprehensive interconnections, and more intelligence. However, while most of the core enabling technologies are general and apply to many application domains, others remain specific to the urban application domain, specifically to the special requirements and objectives of smart sustainable cities.

PERVASIVE SENSING FOR URBAN SUSTAINABILITY

In the emerging field of smart sustainable urban planning, many scholars in different disciplines and practitioners in different professional domains advocate particularly the inclusion of ubiquitous sensing. Sensor ubiquity is a core feature of smart sustainable cities of the future, which rely on the fulfillment of the prevalent ICT visions of pervasive computing. Within the next 15 years or so, most of the data that will be used to monitor, understand, analyze, and plan the systems of smart sustainable cities will come from digital sensing in the form of observations, transactions, and movements associated with the operating and organizing processes of urban life.

Sensors can provide readings on many environmental, social, economic, and physical phenomena. The related data will be available in various forms, with temporal tags and geotags, coupled with a variety of data mining methods and data visualization techniques for displaying and presenting patterns and correlations. A large number of methods for collecting and capturing urban big data from new varieties of digital access are being fashioned and deployed across urban environments. Examples of digital access include the satellite-enabled GPS in vehicles and on citizens, traces left from online transactions processing and related demand-supply situations, online interactions (e.g., social media sites), numerous kinds of websites, and online interactive data systems pertaining to crowdsourcing.

Satellite remote-sensing data are also becoming widely deployed, in addition to a variety of scanning technologies associated with the IoT. The convergence of these phenomena are increasingly paving the way for big data analytics (and context-aware computing) to become the dominant mode of urban analytics in relation to urban operational functioning and planning, as well as for exploiting and extending a variety of data mining and machine learning techniques through which the generation of models will be essential in a wide range of engineering solutions for advancing urban sustainability, i.e., improving the contribution of smart sustainable cities to the goals of sustainable development. Such cities are to be monitored, understood, analyzed, and planned across several spatial levels mostly on the basis of data routinely and automatically collected by sensors.

With the flourishing smart sustainable urban planning approach, pervasive sensing is gaining increased momentum and prevalence as to measuring and collecting data on urban functioning and change in a new way, from the ground up, by means of powerful sensing technologies (motion, behavior, orientation, location, etc.). At present, for instance, sensing urban change from the ground up occurs "through new sensing technologies that depend on hand-held and remote devices through to assembling transactional data from online transactions processing which measure how individuals and groups expend energy, use information, and interact" with respect to resources. Linking and meshing data from various types of sophisticated measuring devices (RFID, NFC, GPS, laser scanners, etc.) with the automation of standard secondary sources of data and unconventional data no doubt provides a rich nexus of possibilities as to providing new and open sources of data necessary for monitoring and understanding how smart sustainable cities will function in a more effective and efficient way.

At present, the urban environment is pervaded by huge quantities of active devices of diverse kinds and forms to particularly automate routine decisions. The fabric of smart sustainable cities is expected to be, arguably, enveloped with an electronic skin, which can be sewed together and entrenched with even more advanced embedded measuring devices, information processing systems, and communication networks. These include countless intelligent sensing and computing devices and related sophisticated and dedicated techniques and algorithms, as well as widespread diffusion of wirelessly ad hoc, mobile network infrastructures, and related protocols. The primary aim is to build an entirely new holistic system which supports the following:

- The acquisition and coordination of data from multiple distributed sources;
- The management and organization of data streams;
- The integration of heterogeneous data into coherent databases and their warehousing;
- The preprocessing and transformation of data;
- The management and seamless composition of extracted models and patterns respectively;
- The evaluation of the quality of the extracted models and patterns;
- The visualization and exploration of behavioral patterns and models;
- The simulation of the mined patterns and models; and
- The deployment of the obtained results for decision support and efficient service provision.

Regardless of their scales, new sensing and computing devices are projected to be equipped with quantum-based processing capacity, unlimited memory size, and high-performance communication capabilities, all linked by mammoth bandwidth and wireless (Internet) connectivity as well as middleware architectures connecting several kinds of distributed, heterogeneous hardware systems and software applications. All of the above is to be directed for advancing the contribution of smart sustainable cities to the goals of sustainable development.

Explicitly, future urban ICT driven by the new wave of computing will result in a blend of advanced applications, services, and computational (data) analytics enabled by constellations of instruments across several spatial scales linked via multiple networks, which can provide a fertile environment conducive to monitoring, understanding, analyzing, evaluating, and planning the sustainability of future cities.

Recent advances in sensor technology have given rise to a new class of miniaturized devices characterized by advanced signal processing methods, high performance, multi-fusion techniques, and high-speed electronic circuits. The trends toward ICT of the new wave of computing, coupled with the evolving concept of smart sustainable cities, are driving research into ever-smaller sizes of sensors capable of powerfully sensing complex and varied aspects of urban life and environment at very low cost. The production of sensing devices with a low cost-to-performance ratio is further driven by the rapid development of sensor manufacturing technologies.

The increasing miniaturization of computer technology is making it possible to develop miniature on-body and remote sensors that allow registering various human and urban parameters without disturbing citizens or interfering with urban activities, thereby the commonsensical infiltration of sensors into daily urban life and environment. This is instrumental in enhancing the computational understanding and data processing of human mobility, urban dynamic processes, and urban operational functioning, a process that entails analysis, interpretation, modeling, and evaluation of big data for enhanced decision-making and deep insights. The new wave of urban computing is about the omnipresence of invisible technology in urban environments and thus citizens' everyday life. Countless tiny, distributed, networked sensor devices will be invisibly embedded in cities for data collection. The research in the area of micro- and nano-engineering is expected to yield major shifts in ICT performance and the way mechatronic components and devices are manufactured, designed, modeled, and implemented, thereby radically changing the nature and structure of sensing devices and thus the way cities will be monitored, understood, analyzed, probed, and planned in the near future.

Automated Forms of Big Data

This chapter is concerned with the automated forms of big data, which is the most useful source of big data, whether enabled by the IoT or other forms of ubiquitous computing, when it comes to city development and planning. Such data are generated as an inherent, automatic function of the devices and systems that are widely deployed and networked across urban environments. Indeed, the automated forms of data generation have recently captured the imagination of those concerned with understanding, managing, and planning cities as well as seeking useful insights into urban systems, in particular in the context of environmental sustainability.

In particular, there has been an interest in sensor networks and the IoT as well as the tracking and tracing of people and objects. Sensor networks involve an array of tiny, inexpensive sensors and actuators that can be embedded or placed on different structures and systems to measure specific physical outputs or properties pertaining to urban environments, including levels of light, humidity, temperature, heat, acoustics, air pressure, motion, orientation, location, speed, and so on. Active sensors broadcast data at regular intervals over local or wide area networks, or might be equipped with NFC capabilities that enable two-way communication.

Sensors networks can be used to monitor the use and condition of public infrastructures, such as buildings, roads, parks, facilities, and utility provision, as well as general environmental conditions within smart sustainable cities. In this context, smart sustainable cities entail a blend of advanced applications, services, and computational (data) analytics enabled by constellations of instruments across many spatial scales linked via multiple networks, which can provide a fertile environment conducive to monitoring, understanding, analyzing, and planning urban systems to improve environmental efficiency and sustainability by providing continuous data regarding physical activities and human movements as well as the status of various forms, structures, and processes. As such, smart sustainable cities offer the prospect of an objectively measured, real-time analysis of the processes operating and organizing urban life, which is of paramount importance to advancing environmental sustainability.

There are a number of different means by which the automated forms of data can be generated, including sensors:

- Capture systems in which the means of performing a task captures data about that task
- Digital traces left through purchase of goods and related demand supply situations
- Digital devices that record and communicate the history of their own use
- Transactions and interactions across digital networks that not only transfer information, but generate data about the transactions and interactions themselves

- Clickstream data that record how people navigate through websites or apps
- Sensed data generated by a variety of sensors and actuators embedded into objects or environments that regularly communicate their measurements
- The scanning of machine-readable objects such as travel passes, passports, or barcodes on parcels that register payment and movement through a system
- Machine to machine interactions across the IoT
- Automatic meter reading (AMR) that communicates utility usage on a continuous basis
- Automated monitoring of public service provision

Uniquely indexical objects and machines that conduct automatic work and are part of the IoT, communicating about their use and traceable if they are mobile (automatic doors, lighting and heating systems, washing machines, security alarms, wifi router boxes, etc.)

Transponders that monitor throughput at toll-booths, measuring vehicle flow along a road or the number of empty spaces in a car park, and track the progress of buses and trains along a route GPS in vehicles and on people RFID tags attached to objects and people Smart tickets that are used to trace passenger travel.

In the domain of urban planning and management, these forms of instrumentation provide abundant, systematic, dynamic, varied, well-defined, resolute, relatively cheap data about urban processes and activities, allowing for real-time analytics and adaptive forms of management and governance. This is of high relevance to environmental sustainability in the context of smart sustainable cities.

SENSOR-BASED URBAN SUSTAINABILITY MINING

As part of urban reality mining (e.g., Batty et al. 2012; Eagle and Peatland 2006), urban sustainability mining, which pertains to sensing complex environmental and socioeconomic systems by means of ubiquitous sensors embedded throughout urban environments, is a key determinant of how cities developing and responding to the challenge of sustainability are becoming smarter. Mining of urban sustainability depends on dedicated, powerful software applications to log urban infrastructures, spatial organizations and interactions, and mobility and travel behavior as well as ecosystem and public services.

The analysis of derived large datasets helps to extract computationally complex activity, behavior, process, and environment models to identify and gain predictive insights into new forms, structures, systems, and processes as to how smart sustainable cities can increase their contribution to sustainability through enhancing urban intelligence functions for decision-making in this regard. Therefore, sensor-based big data have enormous potential to gain new insights into and drive decisions about how sustainability can be better translated into the built, infrastructural, operational, and functional forms of smart sustainable cities across several spatial scales. Further studies in this direction are most likely to enhance mobility, transport engineering, energy engineering, planning, spatial and physical structures, and data-driven characterization of urban functioning in the context of sustainability.

SENSOR TECHNOLOGIES IN CONTEXT-AWARE COMPUTING

Sensor Types and Sensing Areas in Context-Aware Applications

As with big data analytics, context-aware computing involves a wide variety of sensors. A sensor can be described as a device that detects or measures a physical property or some type of input from the physical environment, and then indicates or reacts to it in a particular way. The output is a signal in the form of human-readable display at the sensor location or a recorded data that can be transmitted over a network for further processing. Commonly, sensors can be classified according to the type of energy they detect as signals, and include, but are not limited to, the following types:

- Location sensors (e.g., GPS, active badges);
- Optical/vision sensors (e.g., photodiode, color sensor, IR, and UV sensor);
- Light sensors (e.g., photocells, photodiodes);
- Image sensor (e.g., stereo-type camera, infrared);
- Sound sensors (e.g., microphones);
- Temperature sensors (e.g., thermometers);
- Heat sensors (e.g., bolometer);
- Electrical sensors (e.g., galvanometer);
- Pressure sensors (e.g., barometer, pressure gauges);

Motion sensors (e.g., radar gun, speedometer, mercury switches, tachometer);

- Orientation sensors (e.g., gyroscope);
- Physical movement sensors (e.g., accelerometers);
- Biosensors (e.g., pulse, galvanic skin response measure);
- Vital sign processing devices (heart rate, temperature);
- Wearable sensors (e.g., accelerometers, gyroscopes, magnetometers); and
- Identification and traceability sensors (e.g., RFID, NFC).

While there are different ways of sensing that could be utilized for detecting various features of context, in the realm of smart sustainable cities not all the above are of use in relation to context-aware applications in terms of optimization, control, management, operation, and service delivery associated with sustainability dimensions. How many and what types of sensors can be used in relation to a given context-aware application is determined by the way in which context is operationalized (defined so that it can be technically measured and thus conceptualized) in terms of the number of the entities of context that are to be incorporated in the system based on the application domain, and also whether and how these entities can be combined to generate a high-level abstraction of context (e.g., the physical, situational, behavioral, and social dimension of context). Too often, in relation to both citizens and urban systems, various kinds of sensors are used to detect context.

Acquisition of sensor data about citizens and urban systems (energy, traffic, transport, mobility, etc.) and their behavior and functioning is an important factor in addition to the knowledge domain for analysis of such data by data processing units. In relation to context-aware applications pertaining to citizens, data can be generated from multiple sources, including software equivalents in relation to citizens' devices, such as smartphones, computers, laptops, and other everyday objects. In other words, data are collected and captured from a variety of digital sensors as well as online interactive applications.

Observed information about the citizen and urban system' states or situations in conjunction with the dynamic models for the citizen and system' relevant processes serve as input for the process of computational understanding. This entails the analysis and estimation of what is going on in the surrounding environment in the context of smart sustainable cities. Accordingly, for a context-aware application or system to be able to infer high-level context abstraction based on the interpretation of and reasoning on context information, it is first necessary to acquire low-level data from physical sensors (and other sources). Researchers from different application domains within the field of context-aware computing have investigated context recognition for the past two decades or so by developing a diversity of sensing devices (in addition to methods and techniques for signal and data processing, pattern recognition, modeling, and reasoning tasks). Thus, numerous types of sensors are currently being used to detect various attributes of context.

MULTI-SENSOR DATA FUSION AND ITS APPLICATION IN CONTEXT-AWARE SYSTEMS

In context-aware computing, underlying the multi-sensor fusion methodology is the idea that an abstraction of context as an amalgam of different, interrelated contextual elements can be generated or inferred on the basis of information detected from multiple, heterogeneous data sources, which provide different, yet related, sensor information. Thus, sensors should be integrated to yield optimal context recognition results, i.e., provide a robust estimation of context. A given dimension of the context, a higher level of the context, can be deduced by using a number of external or internal contexts as an atomic level of the context. The use of multi-sensor fusion approach in context-aware applications and systems allows gaining access simultaneously to varied information necessary for accurate estimation or inference of context. Multi-sensor fusion systems have the potential to enhance the information gain while keeping the overall bandwidth low. Figure 1 illustrates a multi-sensor fusion approach.

The use of multi-sensor fusion approach in context-aware applications and systems allows gaining access simultaneously to varied information necessary for accurate estimation or inference of context. Multi-sensor fusion systems have the potential to enhance the information gain while keeping the overall bandwidth low. Figure 1 illustrates a multi-sensor fusion approach.

Wireless Communication Network Technologies and Smart Network Infrastructures

In the context of smart sustainable cities, wireless solutions are set to proliferate in ways that are hard to imagine, as ICT continues to be fast embedded and interwoven into the very fabric of current smart and sustainable cities in terms of their systems and processes in an increasingly computerized urban society. This is a future world of pervasive computing infrastructures and communication networks. Countless sensors will use various wirelessly ad hoc and mobile networks to provide cities with all kinds of data necessary for a wide variety of applications and services. In particular, the widespread diffusion of wireless network technologies will, as a by-product of their normal operations, enable to sense, collect, and coordinate massive repositories of spatiotemporal data pertaining to urban systems, which represent city-wide proxies for all kinds of activities and operating and organizing processes.

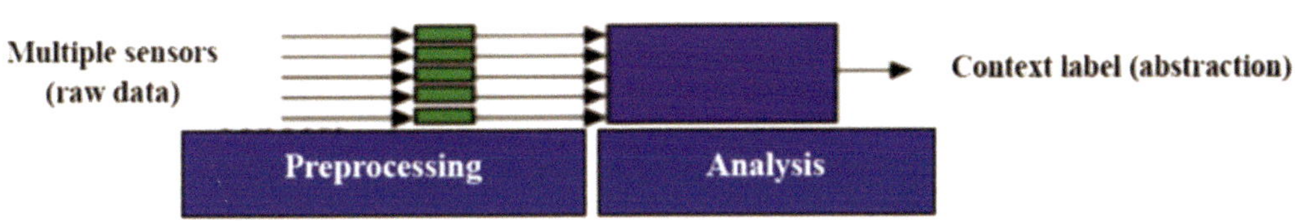

Fig. 1 Use of multiple sensors for context awareness

Also, smart networks are necessary for big data applications in terms of connecting the components and entities of smart sustainable cities, including diverse citizens' everyday objects (computers, smartphones, cars, house devices, etc.) and city infrastructures and facilities as well as urban departments, authorities, and enterprises. Such networks are intended to provide efficient means for transferring the collected data from heterogeneous and distributed sources to data warehouses where big data are to be stored, coalesced, organized, and integrated for processing and analysis in connection with intelligent decision support systems. This involves transferring responses back to the different citizens' devices and urban entities' systems for the purpose of improving different aspects of sustainability.

In relation to ICT of the new wave of computing, networking is a core enabling technology, in addition to cheap, low-power sensing and computing devices. In this context, the role of networking lies in tying hardware and software systems all together for the functioning of ubiquitous applications and services in urban areas, to draw on Bibri.

Accordingly, many heterogeneous components and devices across dispersed infrastructures and disparate networks need to interconnect as part of vast architectures enabling big data analytics, context-aware computing, intelligence functions, and service provisioning on a hard-to-imagine scale. To put it differently, wireless network technologies are prerequisite for coordinating data as well as linking up many diverse distributed sensing devices and computing components and enabling them to interact in the midst of a variety of hardware and software systems necessary for realizing smart urban environments for advancing sustainability.

Wireless technologies, especially satellite-enabled GPS, Wi-Fi, and mobile phone networks enable to sense, collect, and coordinate massive environmental and socioeconomic data representing enormous proxies for the operations, functions, and services of smart sustainable cities and thus powerful physical-environmental and socio-behavioral microscopes. This may facilitate, by means of big data analytics (data mining and database integration capabilities) which offer the prospect for adding value in terms of massive data analysis and integration, discovering the hidden patterns, correlations, and models that characterize, on the one hand, human mobility and movement as part of daily trajectories and activities of citizens and, on the other hand, physical structures and spatial organizations, which can be instrumental in strategic decision-making associated with urban sustainability planning. In all, while pervasive sensing and computing infrastructures allow for monitoring, understanding, and analyzing urban life in terms of infrastructure, built form, administration, and ecosystem and human services, pervasive networking infrastructures allow for collecting and coordinating extensive data in terms of how these data are stored, made accessible, and utilized.

In the context of smart sustainable cities, advanced digital networks are crucial to urban operational functioning and planning due to the interrelationships between urban components and domains that are too many to catalogue (transport, mobility, communication, building, energy, environment, water, waste, land use, healthcare, etc.). These are planned to be further heavily networked while the activities relating to these domains to be linked up. The key domains "which currently are being heavily networked involve: transport systems of all modes in terms of operation, coordination, timetabling, utilities networks which are being enabled using smart metering, local weather, pollution levels and waste disposal, land and planning applications, building technologies in terms of energy and materials, health information systems in terms of access to facilities by patients the list is endless.

The point is that we urgently need a map of this terrain so that we can connect up these diverse activities." Especially, the evolving techno-urban contexts are opening spaces for smart sustainable initiatives in domain networking at current times of tension as alternative trajectories are actively being sought due to the challenge of sustainability, which entails creating innovative solutions that further facilitate collaboration among urban domains and hence integrate urban systems.

In parallel, the aim of emerging technological platforms such as UbiComp, AmI, the IoT, and SenComp is to orchestrate and coordinate the various computational entities in the informational landscape of smart sustainable cities and merging it with their physical landscape into an open system that helps diverse urban entities cope with and plan their activities in relation to improving sustainability. Besides, the growing depth, scale, and complexity of urban networks in terms of both domains and technological infrastructures call for developing and coordinating such networks and enhancing their digital capabilities in ways that increase and sustain the contribution of smart sustainable cities to the goals of sustainable development.

Advanced wireless technologies are extremely placed to initiate this development and coordination. Moreover, with their ever-growing volume, variety, velocity, and timeliness, data on the state of urban networks as built artifacts as well as on that of their use as part of urban activities and processes provide enormous potential to improve urban operational functioning and planning in terms of sustainability, efficiency, and the quality of life by exploiting the analytical power of big data for deep insights and enhanced decision-making. To effectively use these data when implementing big data applications in smart sustainable cities requires fostering these data by advanced wireless technologies, especially in relation to real-time applications. The rationale is that such applications entail that the data from distributed sources should be aggregated and fused prior to being transferred in real-time to cloud computing infrastructures or data processing platforms for stream processing and decision-making. Important to note is that the aggregation and fusion should be carried out in ways that enable data to remain reliable, accurate, and correct for more effective results and thus beneficial knowledge in terms of decision-making processes. This is in turn of critical importance for maintaining the quality and performance of real-time big data applications in terms of decision-making processes.

Data Processing Platforms for Big Data Analytics

There is a variety of available data processing platforms for big data analytics, which provide the stream processing required by real-time big data applications in relation to various urban domains. Therefore, data processing platforms are a key component of the ICT infrastructure of smart sustainable cities of the future with respect to big data applications. Among the leading platforms for big data storage, processing, and management include Hadoop MapReduce, IBM Infosphere Streams, Stratosphere, Spark, and NoSQL—database system management.

These platforms work well on cluster systems to meet the requirements of big data applications for smart sustainable cities; entail scalable, evolvable, optimizable, and reliable software and hardware components; and provide high performance computational and analytical capabilities (namely selection, preprocessing, transformation, mining, evaluation, interpretation, and visualization), in addition to storage, coordination, and management of large datasets across distributed environments. As ecosystems, they perform big data analytics related to a wide variety of large-scale applications intended for different uses associated with the process of sustainable urban development, such as management, control, optimization, assessment, and improvement, thereby spanning a variety of urban domains and sub-domains. In all, they are prerequisite for data-centric applications for smart sustainable cities of the future.

The focus on Hadoop MapReduce is justified by the suitability of its functionalities as to handling urban data as well as to its advantages associated with load balancing, cost effectiveness, flexibility, and processing power compared to other data processing platforms. MapReduce has become the primary big data processing system given its simplicity, scalability, and fine-grain fault tolerance.

Cloud Computing for Big Data Analytics: Characteristic Features and Benefits

Big data analytics can also be performed in the Cloud. This involves both big data Platform as a Service (PaaS) and Infrastructure as a Service (IaaS). Having attracted attention and gained popularity worldwide, cloud computing is becoming increasingly a key part of the ICT infrastructure of both smart cities and sustainable cities as an extension of distributed and grid computing due to the prevalence of sensor technologies, storage facilities, pervasive computing infrastructures, and wireless communication networks.

Especially, most of these technologies have become technically mature and financially affordable by cloud providers. By commoditizing services, low-cost open source software, and geographic distribution, cloud computing is becoming increasingly an attractive option.

Big data analytics is associated with cloud computing, an Internet-based computing model that is increasingly seen as the most suitable solution for highly resource intensive and collaborative applications as an on-demand network access to a shared pool of computing resources (memory capacity, energy, computational power, network bandwidth, interactivity, etc.). This entails that computer-processing resources, which reside in the cloud, are virtualized and dynamic. This implies that only display devices for information and services need to be physically present in relation to urban domains where diverse stakeholders (administrators, planners, landscape architects, sustainability strategists, authorities, citizens, etc.) can make use of software applications and services to improve sustainability.

Such stakeholders can access cloud-based software applications through a web browser and a lean client (a computer program that depends on its server to fulfill its computational roles) or mobile devices while software tools and urban data of all kinds are stored on servers at a remote location. Indeed, cloud computing model is based on hosted services in the sense of application service provisioning running client–server software locally. In this respect, smart sustainable city applications pertaining to transport, traffic, mobility, energy, public health, civil security, education, and so on reside "in the cloud" and can be accessible per demand.

Moreover, the software development platform can be offered in a public, private, or hybrid network, where the cloud provider manages the platform that runs the applications and relieves the cloud clients from the burden of securing dedicated platforms, which would otherwise be very demanding and costly in terms of resources and time. The cloud clients can accordingly benefit from tested, scalable, reliable, and maintainable platforms offered by the cloud provider. Another advantage involves service process optimization through advanced functionalities of software development platforms, namely flexibility, interoperability, reusability, scalability, and cooperation.

There is also a great opportunity to slash or minimize energy consumption associated with the operation of ICT infrastructure, especially when it comes to large-scale deployments like in the case of smart sustainable cities as to different departments and service agencies. Beloglazov develop policies and algorithms that aim at increasing energy efficiency in cloud computing. Energy consumption is way too lower than if all urban entities have their own software development platforms. These are indeed shared by multiple users as well as dynamically reallocated per demand. This approach maximizes the use of computational power and reduces energy usage and thus mitigates GHG emissions associated otherwise with powering a variety of functions as well as data centers dispersed throughout the departments and service agencies of smart sustainable cities.

Whether public or private, the cloud provider includes the cloud environment's servers, storage, networking, and data center operations. This implies that the cloud provider has the actual energy-consuming computational resources; users or clients can simply log on to the network without installing anything, thereby curbing energy usage and making the best of the available computational power. Energy efficiency in cloud computing can result from energy-aware scheduling and server consolidation.

Researchers provide a survey on energy efficiency in cloud computing. Also, cloud computing is seen as a form of green computing, especially if it is based on renewable energy like solar panels. It has other intuitive benefits because it relies on sharing of resources and maximizing the effectiveness of the shared resources, thereby reducing the costs otherwise incurred by ICT operations as to human, technical, and organizational resources. In cloud computing, supercomputers in large data centers as a distributed system of many servers are used to deliver services in a scalable manner as well as to enable the storage and processing of vast quantities of data.

Cloud computing offers great opportunities for streamlining data processing. In all, cloud computing constitutes an efficient and elegant solution in terms of facilitating the huge demand for computing resources associated with big data analytics for decision-making processes in relation to the operational functioning and planning of cities in terms of sustainability. Through the use of cloud computing, smart sustainable cities can accordingly have higher possibilities to perform more effectively and efficiently thanks to the advanced technological features underlying the functioning of cloud computing model.

In addition, cloud computing performs service-oriented computing. In this regard, it can rapidly process large and complex data produced from urban activities and simultaneously serve citizens in relation to healthcare, education, housing, utility, and so on, providing a kind of integrated and specialized center for information services to both the general public and urban departments across various urban domains. In light of this, with reference to smart sustainable cities, cloud computing has the ability to run smart applications on many connected computers and smartphones at the same time for different purposes associated with increasing sustainability performance.

In sum, among the key advantages provided by cloud computing technology include cost reduction, location and device independence, virtualization (sharing of servers and storage devices), multi-tenancy (sharing of costs across a large pool of cloud provider's clients), scalability, performance, reliability, and maintenance. Therefore, opting for cloud computing to perform big data analytics in the realm of smart sustainable cities remains thus far the most suitable option for the operation of infrastructures, applications, and services whose functioning is contingent upon how urban domains interrelate and collaborate, how efficient they are, and to what extent they are scalable as to achieving and maintaining the required level of sustainability.

HOW IOT AND AI REVITALIZE INFRASTRUCTURE SUSTAINABLY

Technology helped Miami-Dade County revitalize part of its critical infrastructure. Consistently delivering water and managing wastewater for millions of people in the seventh-most-populated county in the United States had taken a toll on the Miami-Dade Water and Sewer Department's (WASD) system. The department turned to IoT, a cloud database, and sensors to improve overall operational efficiency.

Cities and county agencies are often overwhelmed by data and how to use it to provide efficient public services. Technologies like IoT and AI can help sift through data to help make smarter decisions that lead to more sustainable communities. Since implementing the solution, WASD has improved service, reduced infrastructure costs, and operates more sustainably.

Smart cities improve overall quality of life

Consider various infrastructure services in your community that maintain your daily quality of life, such as water mains and sewage treatment services. The New York City Department of Environmental Protection oversees the largest combined water and wastewater utility in the U.S., delivering one billion gallons of drinking water every day and treating even more wastewater. The agency's leaders saw potential for using AI-powered chatbots to provide quicker responses for employees calling into the department's IT and human resources desks.

Using services from Microsoft Bot Framework, Skype for Business, SharePoint Online, Microsoft Azure, and Microsoft Dynamics 365, the comprehensive self-service tools for employees can help them reset passwords after business hours, address frequently asked questions, and manage and resolve tickets. As the department sees positive results in how employees are served by the automated chatbots, it plans to expand their use to water and utility customers as well.

Anything we can do to collect more information out of technology will help make better decisions to drive sustainability in the long term and create a better quality of life for everyone.

Smart city technology modernizes healthcare facilities

Human health and well-being are interconnected to the state of one's environment, with clean air and water and quality natural areas providing for our life-sustaining needs. As the largest nonprofit healthcare provider in the greater Seattle area, Swedish Medical Center has reduced its energy consumption and carbon footprint while also achieving regulatory compliance. In partnership with MacDonald Miller Facility Solutions, the organization leveraged Microsoft technology and a partner solution, ICONICS, to modernize its facilities management through preventative maintenance, advanced fault detection, and smart energy analytics. The solution helped to achieve operational sustainability and improve efficiency with $350,000 in energy cost savings during the first year at just one of Swedish's campuses.

Health organizations face unique challenges from disconnected and disparate data. Tracy Picon, Health & Life Sciences Industry Advisor, points to using technology to harness data and provide actionable intelligence to make real-time changes that positively and sustainably impact health organizations and their communities. A healthy society requires a healthy planet.

AI PREPARES SMART CITIES FOR POPULATION GROWTH

Population growth preparation is crucial. Global food demand is expected to grow 41 percent by 2050. Smart agriculture technology, however, can help to increase food production while saving water. Land O' Lakes and Microsoft have partnered to innovate in agriculture, improve the supply chain, and enhance sustainability practices for farmers and the food system.

Addressing population growth and global food demand is imperative, maintains Laura Edell, Chief Data Scientist at Microsoft US. "That's where AI comes into play—creating new solutions can provide insights and intelligence to improve sustainable agriculture practices." Land O' Lakes, one of the largest farmer-owned cooperatives in the U.S. with 150 million acres of productive cropland, is working with Microsoft to ensure farmers have accurate output to meet food demands.

SUSTAINABILITY AS ONE OF THE STRATEGIC GOALS OF SMART CITIES

The steep growth in urban population and the subsequent increase in resource consumption will inevitably create numerous challenges for cities. This fact highlights the importance of shifting paradigms in the way cities work in terms of sustainability. For the purposes of the present study, it is important to establish a working definition of sustainability. Researchers highlight the importance of establishing the scale on which a system is being assessed in terms of sustainability. Achieving sustainability on a global scale requires different type of actions than on an urban scale. There is no single best-established definition in terms of sustainability in the urban scale nevertheless there is a commonly-used set of characteristics of urban sustainability. These include intergenerational equity, intra-generational equity (social, geographical, and governance equity), conservation of the natural environment, significant reduction of the use of non-renewable resources, economic vitality and diversity, autonomy in communities, citizen well-being, and gratification of fundamental human needs.

These characteristics incorporate the three dimensions of sustainability: the environmental, the economic and the social dimension, where the environmental regards the ecological aspect and includes the conservation of the natural environment (flora and fauna) and natural resources and an energy production based economy. The social dimension includes equity, community autonomy, citizen well-being, and gratification of fundamental human needs, while the economic one consists of the economic vitality and diversity of urban areas. For the context of this research an urban environment can be sustainable when social equity, conservation of the natural environment and its resources, economic vitality and quality of life are achieved. Urban sustainability appears to be one of the prevailing themes in smart city literature, but to what extent is the concept embedded in the understanding of smart cities and how comprehensively is it addressed?

THE FUTURE OF SUSTAINABLE URBANISM

On the basis of the reviews of a history of sustainable urbanism and the literature on green cities, or eco-cities, six general principles for designing a sustainable city are defined:

- Close cycles at the lowest possible scale;
- Build redundancy in the design;
- Create anti-fragility;
- See citizens as (design) experts;
- Use the landscape as the basis for urban growth; and
- Develop innovative, rule-breaking designs.

Bringing the analyses of the history of sustainability in urbanism and the theory of eco-cities together, three comprehensive aspects of sustainable urbanism can be distinguished: the relation between society and the research community, the complexity of urban resilience, and the role of 'landscape' in urbanism.

PRINCIPLES FOR SUSTAINABLE URBANISM

The six principles for sustainable urbanism are the following:

- **Close cycles at the lowest possible scale.** When the cycles of water, energy, and materials are closed, no waste streams are produced and the city is clean and has an efficient urban metabolism. The circularity provides abundant opportunities to create economic benefits. However, it is not always technically and financially most feasible to close cycles at the lowest possible scale. As Tillie and Van den Dobbelsteen have pointed out, for energy solutions on different scales, up-scaling from the individual building can also lead to a balanced urban system;

- **Create spaces for the unknown.** When redundancy becomes a standard part in urban design, the city has the flexibility to 'host' unprecedented events. If, for instance, a cyclone hits the city, there is space to temporarily store the accompanying water. Roggema has identified this space as being 'unplanned';
- **Design anti-fragile spaces.** Using this new concept to design places that improve the quality in the city when an event occurs. 'Ant-fragility is defined as a convex response to a stressor or source of harm (for some range of variation), leading to a positive sensitivity to increase in volatility (or variability, stress, dispersion of outcomes, or uncertainty, what is grouped under the designation "disorder cluster")'. Applied in urban environments under stress of climate impacts, anti-fragility is the concept that urban environments become stronger than before through their response to climate events. In urbanism, this concept is ill-researched. However, it could potentially open new design pathways to create urban environments that not only could withstand or bounce back from climate disasters, but, as a result of the hazard, grow stronger;
- **Let people own their environment.** When residents are part of the design process, they will create ownership over the changes and invest in their environment. Once residents own their environment, they will make sure it is maintained well. The design charrette is a wonderful design method to develop designs in an inclusive way.
- **Use the landscape as the basis for urbanism.** Landscape systems, such as the water system, ecology, and soil, create the conditions for urban occupancy. The values in the landscape then will determine the type of urban development. This is contrary to the current situation where the urban development dictates the landscape, e.g., overthrows the existing landscape values to create ordinary residential areas. Even in existing cities, the underlying landscape can be illuminated and brought to the surface to give the city extra ecological qualities.
- **Create innovative designs.** The current conventions need to be broken if a sustainable city is to be developed. The existing procedures need to be bypassed using creative design approaches. For the development of regional plans, metropolitan spatial visions, district plans, and neighbourhood designs, a sabbatical detour can provide the space to think differently and the time to come up with alternative design propositions. Elements of a sabbatical detour are design competitions, an idea-generating charrette, designers in resident, exhibitions, and many other activities.

entire metropolis. Because the set of 15 layers contains technical, socio-economic, and nature-based aspects, the combination of these layers provides a balanced and integrated way of looking at the entire city.

Research on these 15 layers is elaborate and the set-up of a rigorous data-base is time-consuming. Recently, investigations into the Zero-Carbon Metropolis of Sydney has led to an image of the first layer (Figure 1), which already can form the basis for further resilient designs at a lower scale [100]. This first layer has been developed by mapping the metropolitan renewable energy resources of the Greater Sydney area. These mappings of the wind potentials, hydropower potentials, solar potentials, and biomass potentials have been connected with the demand of households, public utilities, and industries under an ecological 'zero-carbon' lifestyle scenario. The integrated map shows the combination of these types of renewable energy supplies.

Landscape-Driven Design

Landscape is everywhere around us, but in recent decades it has not played a dominant role in the growth of cities. The landscape is often neglected and replaced by an urban landscape that marginally contains elements of the former landscape. Both reviews illustrate that the landscape, represented through its elevation, water, ecology, is a crucial factor to create more sustainable cities. These factors, if not used as a determining factor in urban development, will eventually counterpunch at the city in the form of severe flooding, absence of agricultural fertility, or other disruptive processes. When these factors, however, are used as the basis for urban development, a truly sustainable city could emerge.

The processes that drive the landscape are then also the drivers of urban development. If the flows of water form the basis for the urban lay-out, or if the connected habitats of natural ecologies are taken as the starting point of urban development, or if the natural gradients of higher oligotrophic grounds and the lower eutrophic soils are kept intact, the city that emerges builds on the natural richness of the landscape, creating a city in harmony with its environments. Recently, studies have been undertaken for the Badgerys Creek area, west of Sydney, where a new airport and a city with one million new inhabitants are planned to arise. If the urban growth is guided by the principles of the landscape, a very interesting and attractive urban environment can be designed.

CHAPTER

11 DESIGN AND CONSTRUCTION AS AN INTEGRATED SYSTEM

In the planning of facilities, it is important to recognize the close relationship between design and construction. These processes can best be viewed as an integrated system. Broadly speaking, design is a process of creating the description of a new facility, usually represented by detailed plans and specifications; construction planning is a process of identifying activities and resources required to make the design a physical reality. Hence, construction is the implementation of a design envisioned by architects and engineers. In both design and construction, numerous operational tasks must be performed with a variety of precedence and other relationships among the different tasks.

Several characteristics are unique to the planning of constructed facilities and should be kept in mind even at the very early stage of the project life cycle. These include the following:

- Nearly every facility is custom designed and constructed, and often requires a long time to complete.
- Both the design and construction of a facility must satisfy the conditions peculiar to a specific site.
- Because each project is site specific, its execution is influenced by natural, social and other locational conditions such as weather, labor supply, local building codes, etc.
- Since the service life of a facility is long, the anticipation of future requirements is inherently difficult.
- Because of technological complexity and market demands, changes of design plans during construction are not uncommon.

In an integrated system, the planning for both design and construction can proceed almost simultaneously, examining various alternatives which are desirable from both viewpoints and thus eliminating the necessity of extensive revisions under the guise of value engineering. Furthermore, the review of designs with regard to their constructibility can be carried out as the project progresses from planning to design. For example, if the sequence of assembly of a structure and the critical loadings on the partially assembled structure during construction are carefully considered as a part of the overall structural design, the impacts of the design on construction falsework and on assembly details can be anticipated. However, if the design professionals are expected to assume such responsibilities, they must be rewarded for sharing the risks as well as for undertaking these additional tasks.

Similarly, when construction contractors are expected to take over the responsibilities of engineers, such as devising a very elaborate scheme to erect an unconventional structure, they too must be rewarded accordingly. As long as the owner does not assume the responsibility for resolving this risk-reward dilemma, the concept of a truly integrated system for design and construction cannot be realized.

It is interesting to note that European owners are generally more open to new technologies and to share risks with designers and contractors. In particular, they are more willing to accept responsibilities for the unforeseen subsurface conditions in geotechnical engineering. Consequently, the designers and contractors are also more willing to introduce new techniques in order to reduce the time and cost of construction. In European practice, owners typically present contractors with a conceptual design, and contractors prepare detailed designs, which are checked by the owner's engineers. Those detailed designs may be alternate designs, and specialty contractors may also prepare detailed alternate designs.The willingness to assume responsibilities does not come easily from any party in the current litigious climate of the construction industry in the United States. On the other hand, if owner, architect, engineer, contractor and other groups that represent parts of the industry do not jointly fix the responsibilities of various tasks to appropriate parties, the standards of practice will eventually be set by court decisions. In an attempt to provide a guide to the entire spectrum of participants in a construction project, the American Society of Civil Engineers issued a Manual of Professional Practice entitled *Quality in the Constructed Project* in 1990. This manual is intended to help bring a turn around of the fragmentation of activities in the design and construction process.

Shop drawings represent the assembly details for erecting a structure which should reflect the intent and rationale of the original structural design. They are prepared by the construction contractor and reviewed by the design professional. However, since the responsibility for preparing shop drawings was traditionally assigned to construction contractors, design professionals took the view that the review process was advisory and assumed no responsibility for their accuracy. This justification was ruled unacceptable by a court in connection with the walkway failure at the Hyatt Hotel in Kansas City in 1985. In preparing the ASCE Manual of

Professional Practice for Quality in the Constructed Project, the responsibilities for preparation of shop drawings proved to be the most difficult to develop.

Traditionally, the owner is not involved in the preparation and review of shop drawings, and perhaps is even unaware of any potential problems. In the recommended practice, the owner is required to take responsibility for providing adequate time and funding, including approval of scheduling, in order to allow the design professionals and construction contractors to perform satisfactorily.

MODEL METRO PROJECT IN MILAN, ITALY

Under Italian law, unforeseen subsurface conditions are the owner's responsibility, not the contractor's. This is a striking difference from U.S. construction practice where changed conditions clauses and claims and the adequacy of prebid site investigations are points of contention. In effect, the Italian law means that the owner assumes those risks. But under the same law, a contractor may elect to assume the risks in order to lower the bid price and thereby beat the competition.

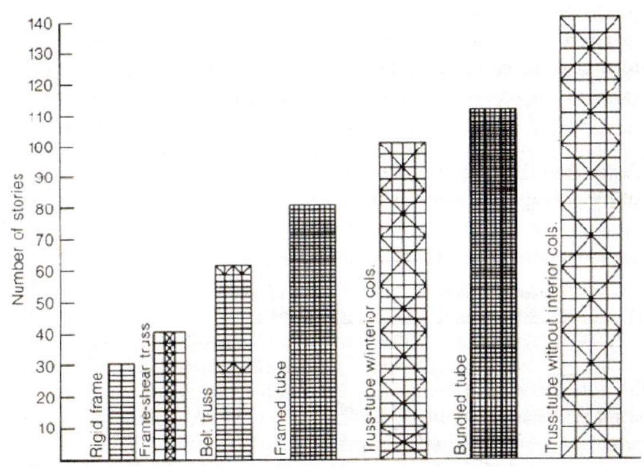

Fig. 1: Proposed Structural System fir Steel Buildings

According to the Technical Director of Rodio, the Milan-based contractor which is heavily involved in the grouting job for tunneling in the Model Metro project in Milan, Italy, there are two typical contractual arrangements for specialized subcontractor firms such as theirs. One is to work on a unit price basis with no responsibility for the design. The other is what he calls the "nominated subcontractor" or turnkey method: prequalified subcontractors offer their own designs and guarantee the price, quality, quantities, and, if they wish, the risks of unforeseen conditions.

At the beginning of the Milan metro project, the Rodio contract ratio was 50/50 unit price and turnkey. The firm convinced the metro owners that they would save money with the turnkey approach, and the ratio became 80% turnkey. What's more, in the work packages where Rodio worked with other grouting specialists, those subcontractors paid Rodio a fee to assume all risks for unforeseen conditions.

Under these circumstances, it was critical that the firm should know the subsurface conditions as precisely as possible, which was a major reason why the firm developed a computerized electronic sensing program to predict stratigraphy and thus control grout mixes, pressures and, most important, quantities.

INNOVATION AND TECHNOLOGICAL FEASIBILITY

The planning for a construction project begins with the generation of concepts for a facility which will meet market demands and owner needs. Innovative concepts in design are highly valued not for their own sake but for their contributions to reducing costs and to the improvement of aesthetics, comfort or convenience as embodied in a well-designed facility.

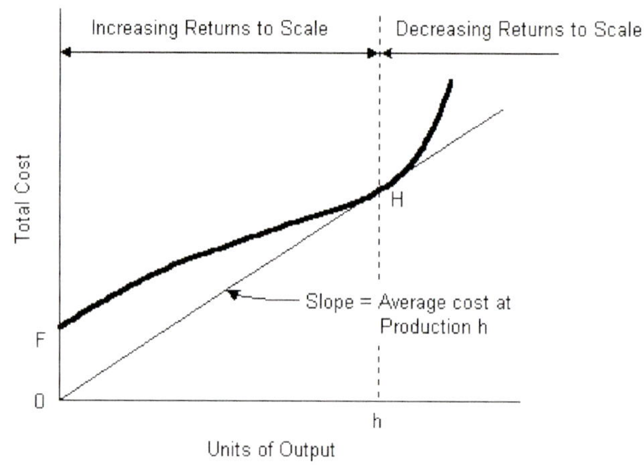

Fig. 2: Market Demand and Total Cost Relationship

However, the constructor as well as the design professionals must have an appreciation and full understanding of the technological complexities often associated with innovative designs in order to provide a safe and sound facility. Since these concepts are often preliminary or tentative, screening studies are carried out to determine the overall technological viability and economic attractiveness without pursuing these concepts in great detail. Because of the ambiguity of the objectives and the uncertainty of external events, screening studies call for uninhibited innovation in creating new concepts and judicious judgment in selecting the appropriate ones for further consideration.

One of the most important aspects of design innovation is the necessity of communication in the design/construction partnership. In the case of bridge design, it can be illustrated by the following quotation from Lin and Gerwick concerning bridge construction:

The great pioneering steel bridges of the United States were built by an open or covert alliance between designers and constructors. The turnkey approach of designer-constructor has developed and built our chemical plants, refineries, steel plants, and nuclear power plants. It is time to ask, seriously, whether we may not have adopted a restrictive approach by divorcing engineering and construction in the field of bridge construction.

If a contractor-engineer, by some stroke of genius, were to present to design engineers today a wonderful new scheme for long span prestressed concrete bridges that made them far cheaper, he would have to make these ideas available to all other constructors, even limiting or watering them down so as to "get a group of truly competitive bidders." The engineer would have to make sure that he found other contractors to bid against the ingenious innovator.

If an engineer should, by a similar stroke of genius, hit on such a unique and brilliant scheme, he would have to worry, wondering if the low bidder would be one who had any concept of what he was trying to accomplish or was in any way qualified for high class technical work.

Innovative design concepts must be tested for technological feasibility. Three levels of technology are of special concern: technological requirements for operation or production, design resources and construction technology. The first refers to the new technologies that may be introduced in a facility which is used for a certain type of production such as chemical processing or nuclear power generation. The second refers to the design capabilities that are available to the designers, such as new computational methods or new materials. The third refers to new technologies which can be adopted to construct the facility, such as new equipment or new construction methods.

A new facility may involve complex new technology for operation in hostile environments such as severe climate or restricted accessibility. Large projects with unprecedented demands for resources such as labor supply, material and infrastructure may also call for careful technological feasibility studies. Major elements in a feasibility study on production technology should include, but are not limited to, the following:

- Project type as characterized by the technology required, such as synthetic fuels, petrochemicals, nuclear power plants, etc.
- Project size in dollars, design engineer's hours, construction labor hours, etc.
- Design, including sources of any special technology which require licensing agreements.
- Project location which may pose problems in environmental protection, labor productivity and special risks.

An example of innovative design for operation and production is the use of entropy concepts for the design of integrated chemical processes. Simple calculations can be used to indicate the minimum energy requirements and the least number of heat exchange units to achieve desired objectives. The result is a new incentive and criterion for designers to achieve more effective designs. Numerous applications of the new methodology has shown its efficacy in reducing both energy costs and construction expenditures. This is a case in which innovative design is not a matter of trading-off operating and capital costs, but better designs can simultaneously achieve improvements in both objectives.

The choice of construction technology and method involves both *strategic* and *tactical* decisions about appropriate technologies and the best sequencing of operations. For example, the extent to which prefabricated facility components will be used represents a *strategic* construction decision. In turn, prefabrication of components might be accomplished off-site in existing manufacturing facilities or a temporary, on-site fabrication plant might be used.

Another example of a strategic decision is whether to install mechanical equipment in place early in the construction process or at an intermediate stage. Strategic decisions of this sort should be integrated with the process of facility design in many cases. At the tactical level, detailed decisions about how to accomplish particular tasks are required, and such decisions can often be made in the field.

Construction planning should be a major concern in the development of facility designs, in the preparation of cost estimates, and in forming bids by contractors. Unfortunately, planning for the construction of a facility is often treated as an after thought by design professionals. This contrasts with manufacturing practices in which the *assembly* of devices is a major concern in design. Design to insure ease of assembly or construction should be a major concern of engineers and architects. As the Business Roundtable noted, "All too often chances to cut schedule time and costs are lost because construction operates as a production process separated by a chasm from financial planning, scheduling, and engineering or architectural design. Too many engineers, separated from field experience, are not up to date about how to build what they design, or how to design so structures and equipment can be erected most efficiently."

Innovative use of structural frames for buildings

The structural design of skyscrapers offers an example of innovation in overcoming the barrier of high costs for tall buildings by making use of new design capabilities. A revolutionary concept in skyscraper design was introduced in the 1960's by Fazlur Khan who argued that, for a building of a given height, there is an appropriate structural system which would produce the most efficient use of the material.

Before 1965, most skyscrapers were steel rigid frames. However, Fazlur Khan believed that it was uneconomical to construct all office buildings of rigid frames, and proposed an array of appropriate structural systems for steel buildings of specified heights as shown in Fig. 1.

By choosing an appropriate structural system, an engineer can use structural materials more efficiently. For example, the 60-story Chase Manhattan Building in New York used about 60 pounds per square foot of steel in its rigid frame structure, while the 100-story John Hancock Center in Chicago used only 30 pounds per square foot for a trusted tube system.

At the time the Chase Manhattan Building was constructed, no bracing was used to stiffen the core of a rigid frame building because design engineers did not have the computing tools to do the complex mathematical analysis associated with core bracing.

INNOVATION AND ECONOMIC FEASIBILITY

Innovation is often regarded as the engine which can introduce construction economies and advance labor productivity. This is obviously true for certain types of innovations in industrial production technologies, design capabilities, and construction equipment and methods. However, there are also limitations due to the economic infeasibility of such innovations, particularly in the segments of construction industry which are more fragmented and permit ease of entry, as in the construction of residential housing.

Market demand and firm size play an important role in this regard. If a builder is to construct a larger number of similar units of buildings, the cost per unit may be reduced. This relationship between the market demand and the total cost of production may be illustrated schematically as in Fig. 2. An initial threshold or fixed cost F is incurred to allow any production. Beyond this threshold cost, total cost increases faster than the units of output but at a decreasing rate. At each point on this total cost curve, the average cost is represented by the slope of a line from the origin to the point on the curve. At a point H, the average cost per unit is at a minimum.

Beyond H to the right, the total cost again increases faster than the units of output and at an increasing rate. When the rate of change of the average cost slope is decreasing or constant as between 0 and H on the curve, the range between 0 and H is said to be *increasing return to scale*; when the rate of change of the average cost slope is increasing as beyond H to the right, the region is said to be *decreasing return to scale*. Thus, if fewer than h units are constructed, the unit price will be higher than that of exactly h units. On the other hand, the unit price will increase again if more than h units are constructed.

Nowhere is the effect of market demand and total cost more evident than in residential housing. The housing segment in the last few decades accepted many innovative technical improvements in building materials which were promoted by material suppliers. Since material suppliers provide products to a large number of homebuilders and others, they are in a better position to exploit production economies of scale and to support new product development. However, homebuilders themselves have not been as successful in making the most fundamental form of innovation which encompasses changes in the technological process of homebuilding by shifting the mixture of labor and material inputs, such as substituting large scale off-site prefabrication for on-site assembly.

There are several major barriers to innovation in the technological process of homebuilding, including demand instability, industrial fragmentation, and building codes. Since market demand for new homes follows demographic trends and other socio-economic conditions, the variation in home building has been anything but regular. The profitability of the homebuilding industry has closely matched aggregate output levels. Since entry and exist from the industry are relatively easy, it is not uncommon during periods of slack demand to find builders leaving the market or suspending their operations until better times. The inconsistent levels of retained earnings over a period of years, even among the more established builders, are likely to discourage support for research and development efforts which are required to nurture innovation.

Furthermore, because the homebuilding industry is fragmented with a vast majority of homebuilders active only in local regions, the typical homebuilder finds it excessively expensive to experiment with new designs. The potential costs of a failure or even a moderately successful innovation would outweigh the expected benefits of all but the most successful innovations. Variation in local building codes has also caused inefficiencies although repeated attempts have been made to standardize building codes.

In addition to the scale economies visible within a sector of the construction market, there are also possibilities for scale economies in individual facility. For example, the relationship between the size of a building (expressed in square feet) and the input labor (expressed in laborhours per square foot) varies for different types and sizes of buildings.

Use of new materials

In recent years, an almost entirely new set of materials is emerging for construction, largely from the aerospace and electronics industries. These materials were developed from new knowledge about the structure and properties of materials as well as new techniques for altering existing materials. Additives to traditional materials such as concrete and steel are particularly prominent. For example, it has been known for some time that polymers would increase concrete strength, water resistance and ability to insulate when they are added to the cement. However, their use has been limited by their costs since they have had to replace as much as 10 percent of the cement to be effective. However, Swedish researchers have helped reduce costs by using polymer microspheres 8 millionths of an inch across, which occupy less than one percent of the cement. Concretes made with these microspheres meet even the strict standards for offshore structures in the North Sea. Research on micro-additives will probably produce useful concretes for repairing road and bridges as well.

Green Buildings

The Leadership in Energy and Environmental Design (LEED) Green Building Rating System is intended to promote voluntary improvements in design and construction practices. In the rating system, buildings receive points for a variety of aspects, including reduced energy use, greater use of daylight rather than artificial lights, recycling construction waste, rainfall runoff reduction, availability of public transit access, etc. If a building accumulates a sufficient number of points, it may be certified by the Green Building Alliance as a "green building." While some of these aspects may increase construction costs, many reduce operating costs or make buildings more attractive. Green building approaches are spreading to industrial plants and other types of construction.

DESIGN METHODOLOGY

While the conceptual design process may be formal or informal, it can be characterized by a series of actions: formulation, analysis, search, decision, specification, and modification. However, at the early stage in the development of a new project, these actions are highly interactive as illustrated in Fig. 3. Many iterations of redesign are expected to refine the functional requirements, design concepts and financial constraints, even though the analytic tools applied to the solution of the problem at this stage may be very crude.

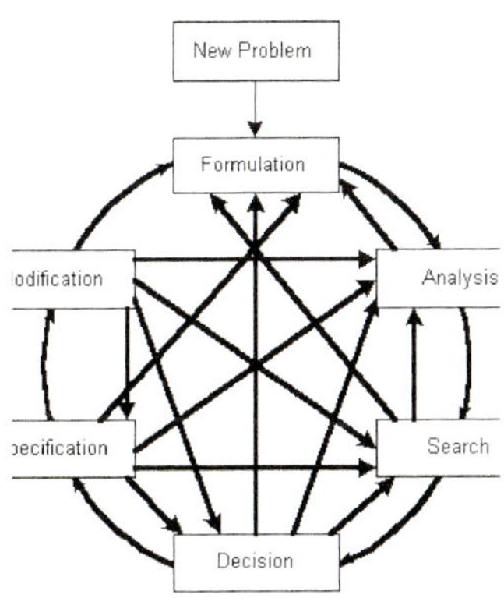

Fig. 3: Conceptual Design Process

The series of actions taken in the conceptual design process may be described as follows:

- Formulation refers to the definition or description of a design problem in broad terms through the synthesis of ideas describing alternative facilities.
- Analysis refines the problem definition or description by separating important from peripheral information and by pulling together the essential detail. Interpretation and prediction are usually required as part of the analysis.
- Search involves gathering a set of potential solutions for performing the specified functions and satisfying the user requirements.
- Decision means that each of the potential solutions is evaluated and compared to the alternatives until the best solution is obtained.
- Specification is to describe the chosen solution in a form which contains enough detail for implementation.
- Modification refers to the change in the solution or re-design if the solution is found to be wanting or if new information is discovered in the process of design.

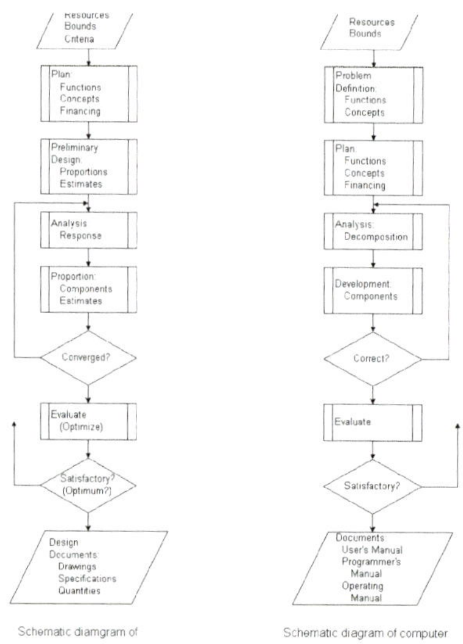

Fig. 4: An Analogy Between Structural Design and Computer Program Development Process

As the project moves from conceptual planning to detailed design, the design process becomes more formal. In general, the actions of formulation, analysis, search, decision, specification and modification still hold, but they represent specific steps with less random interactions in detailed design. The design methodology thus formalized can be applied to a variety of design problems. For example, the analogy of the schematic diagrams of the structural design process and of the computer program development process is shown in Fig. 4.

The basic approach to design relies on decomposition and integration. Since design problems are large and complex, they have to be decomposed to yield subproblems that are small enough to solve. There are numerous alternative ways to decompose design problems, such as decomposition by functions of the facility, by spatial locations of its parts, or by links of various functions or parts. Solutions to subproblems must be integrated into an overall solution. The integration often creates conceptual conflicts which must be identified and corrected. A hierarchical structure with an appropriate number of levels may be used for the decomposition of a design problem to subproblems. For example, in the structural design of a multistory building, the building may be decomposed into floors, and each floor may in turn be decomposed into separate areas. Thus, a hierarchy representing the levels of building, floor and area is formed.

Different design styles may be used. The adoption of a particular style often depends on factors such as time pressure or available design tools, as well as the nature of the design problem. Examples of different styles are:

- **Top-down design.** Begin with a behavior description of the facility and work towards descriptions of its components and their interconnections.
- **Bottom-up design.** Begin with a set of components, and see if they can be arranged to meet the behavior description of the facility.

The design of a new facility often begins with the search of the files for a design that comes as close as possible to the one needed. The design process is guided by accumulated experience and intuition in the form of heuristic rules to find acceptable solutions. As more experience is gained for this particular type of facility, it often becomes evident that parts of the design problem are amenable to rigorous definition and algorithmic solution. Even formal optimization methods may be applied to some parts of the problem.

FUNCTIONAL DESIGN

The objective of functional design for a proposed facility is to treat the facility as a complex system of interrelated spaces which are organized systematically according to the functions to be performed in these spaces in order to serve a collection of needs. The arrangement of physical spaces can be viewed as an iterative design process to find a suitable floor plan to facilitate the movement of people and goods associated with the operations intended.

A designer often relies on a heuristic approach, i.e., applying selected rules or strategies serving to stimulate the investigation in search for a solution. The heuristic approach used in arranging spatial layouts for facilities is based generally on the following considerations:

- Identification of the goals and constraints for specified tasks,
- Determination of the current state of each task in the iterative design process,
- Evaluation of the differences between the current state and the goals,
- Means of directing the efforts of search towards the goals on the basis of past experience.

Hence, the procedure for seeking the goals can be recycled iteratively in order to make tradeoffs and thus improve the solution of spatial layouts.

Consider, for example, an integrated functional design for a proposed hospital. Since the responsibilities for satisfying various needs in a hospital are divided among different groups of personnel within the hospital administrative structure, a hierarchy of functions corresponding to different levels of responsibilities is proposed in the systematic organization of hospital functions. In this model, the functions of a hospital system are decomposed into a hierarchy of several levels:

- **Hospital**—conglomerate of all hospital services resulting from top policy decisions,
- **Division**—broadly related activities assigned to the same general area by administrative decisions,
- **Department**—combination of services delivered by a service or treatment group,
- **Suite**—specific style of common services or treatments performed in the same suite of rooms,
- **Room**—all activities that can be carried out in the same internal environment surrounded by physical barriers,
- **Zone**—several closely related activities that are undertaken by individuals,
- **Object**—a single activity associated with an individual.

In the integrated functional design of hospitals, the connection between physical spaces and functions is most easily made at the lowest level of the hierarchy, and then extended upward to the next higher level. For example, a bed is a physical object immediately related to the activity of a patient. A set of furniture consisting of a bed, a night table and an armchair arranged comfortably in a zone indicates the sphere of private activities for a patient in a room with multiple occupancy. Thus, the spatial representation of a hospital can be organized in stages starting from the lowest level and moving to the top. In each step of the organization process, an element (space or function) under consideration can be related directly to the elements at the levels above it, to those at the levels below it, and to those within the same level.

Since the primary factor relating spaces is the movement of people and supplies, the objective of arranging spaces is the minimization of movement within the hospital. On the other hand, the internal environmental factors such as atmospheric conditions (pressure, temperature, relative humidity, odor and particle pollution), sound, light and fire protection produce constraining effects on the arrangement of spaces since certain spaces cannot be placed adjacent to other spaces because of different requirements in environmental conditions.

The consideration of logistics is important at all levels of the hospital system. For example, the travel patterns between objects in a zone or those between zones in a room are frequently equally important for devising an effective design. On the other hand, the adjacency desirability matrix based upon environmental conditions will not be important for organization of functional elements below the room level since a room is the lowest level that can provide a physical barrier to contain desirable environmental conditions.

Hence, the organization of functions for a new hospital can be carried out through an interactive process, starting from the functional elements at the lowest level that is regarded as stable by the designer, and moving step by step up to the top level of the hierarchy. Due to the strong correlation between functions and the physical spaces in which they are performed, the arrangement of physical spaces for accommodating the functions will also follow the same iterative process. Once a satisfactory spatial arrangement is achieved, the hospital design is completed by the selection of suitable building components which complement the spatial arrangement.

Top-down design style

In the functional design of a hospital, the designer may begin with a "reference model", i.e. the spatial layouts of existing hospitals of similar size and service requirements. On the basis of past experience, spaces are allocated to various divisions as shown schematically in Fig. 5.

The space in each division is then divided further for various departments in the division, and all the way down the line of the hierarchy. In every step along the way, the pertinent information of the elements immediately below the level under consideration will be assessed in order to provide input for making necessary adjustments at the current level if necessary. The major drawback of the top-down design style is that the connection between physical spaces and functions at lower levels cannot be easily anticipated. Consequently, the new design is essentially based on the intuition and experience of the designer rather than an objective analysis of the functions and space needs of the facility. Its greatest attraction is its simplicity which keeps the time and cost of design relatively low.

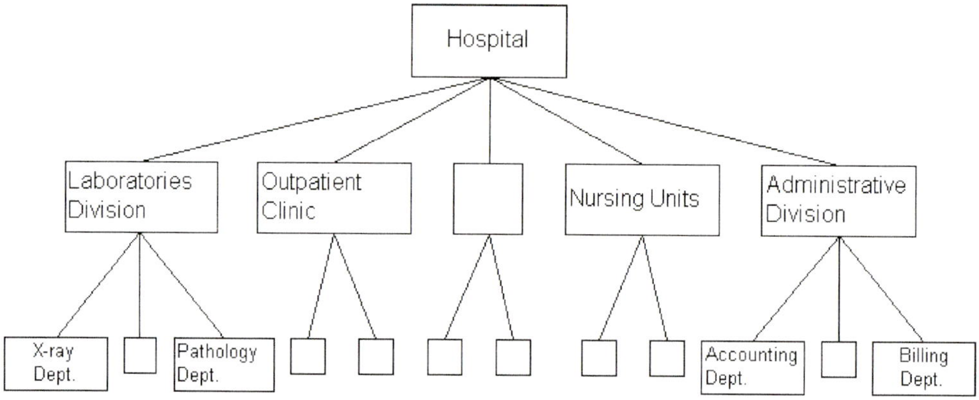

Fig. 5: A Model for Top-Down Design of a Hospital

Bottom-up design style

A multi-purpose examination suite in a hospital is used as an illustration of bottom-up design style. Thus the size of the room is determined by spatial layout required to perform the desired services. Finally, the suite is defined by the rooms which are parts of the multi-purpose examination suite.

Physical Structures

The structural design of complex engineering systems generally involves both synthesis and analysis. Synthesis is an inductive process while analysis is a deductive process. The activities in synthesis are often described as an art rather than a science, and are regarded more akin to creativity than to knowledge. The conception of a new structural system is by and large a matter of subjective decision since there is no established procedure for generating innovative and highly successful alternatives. The initial selection of a workable system from numerous possible alternatives relies heavily on the judicious judgment of the designer. Once a structural system is selected, it must be subjected to vigorous analysis to insure that it can sustain the demands in its environment. In addition, compatibility of the structural system with mechanical equipment and piping must be assured.

For traditional types of structures such as office buildings, there are standard systems derived from the past experience of many designers. However, in many situations, special systems must be developed to meet the specified requirements. The choice of materials for a structure depends not only on the suitability of materials and their influence on the form of the structure. For example, in the design of an airplane hangar, a steel skeleton frame may be selected because a similar frame in reinforced concrete will limit the span of the structure owing to its unfavorable ratio or resistance to weight. However, if a thin-shelled roof is adopted, reinforced concrete may prove to be more suitable than steel. Thus, the interplay of the structural forms and materials affects the selection of a structural system, which in turn may influence the method of construction including the use of falsework.

Steel frame supporting a turbo-blower

The design of a structural frame supporting a turbo-blower supplying pressurized air to a blast furnace in a steel mill can be used to illustrate the structural design process. Since the vibration of the turbo-blower is a major concern to its operation, a preliminary investigation calls for a supporting frame which is separated from the structural frame of the building. An analysis of the vibration characteristics of the turbo-blower indicates that the lowest mode of vibration consists of independent vibration of the turbine shaft and the blower shaft, with higher modes for the coupled turbo-blower system when both shafts vibrate either in-phase or out-of-phase. Consequently, a steel frame with separate units for the blower side and the turbine side is selected. The columns of the steel frame are mounted on pile foundation and all joints of the steel frame are welded to reduce the vibration levels.

Since the structural steel frame also supports a condenser, an air inlet and exhaust, and a steam inlet and exhaust in addition to the turbo-blower, a static analysis is made to size its members to support all applied loads. Then, a dynamic analysis is conducted to determine the vibration characteristics of the system incorporating the structural steel frame and the turbo-blower. When the limiting conditions for static loads and natural frequencies of vibration are met, the design is accepted as satisfactory.

Multiple hierarchy descriptions of projects

In the previous section, a hierarchy of functional spaces was suggested for describing a facility. This description is appropriate for functional design of spaces and processes within a building, but may be inadequate as a view of the facility's structural systems. A hierarchy suitable for this purpose might divide elements into *structural functions* such as slabs, walls, frames, footings, piles or mats. Lower levels of the hierarchy would describe individual design elements. For example, frames would be made up of column, beam and diagonal groups which, in turn, are composed of individual structural elements. These individual structural elements comprise the limits on functional spaces such as rooms in a different hierarchical perspective. Designers typically will initiate a view appropriate for their own concerns, and these different hierarchical views must be synthesized to insure consistency and adequacy of the overall design.

Geotechnical Engineering Investigation

Since construction is site specific, it is very important to investigate the subsurface conditions which often influence the design of a facility as well as its foundation. The uncertainty in the design is particularly acute in geotechnical engineering so that the assignment of risks in this area should be a major concern. Since the degree of uncertainty in a project is perceived differently by different parties involved in a project, the assignment of unquantifiable risks arising from numerous unknowns to the owner, engineer and contractor is inherently difficult. It is no wonder that courts or arbitrators are often asked to distribute equitably a risk to parties who do not perceive the same risks and do not want to assume a disproportionate share of such risks.

Design of a tie-back retaining wall

This example describes the use of a tie-back retaining wall built in the 1960's when such construction was uncommon and posed a considerable risk. The engineer designing it and the owner were aware of the risk because of potentially extreme financial losses from both remedial and litigation costs in the event that the retaining wall failed and permitted a failure of the slope. But the benefits were perceived as being worth the risk—benefits to the owner in terms of both lower cost and shorter schedule, and benefits to the engineer in terms of professional satisfaction in meeting the owner's needs and solving what appeared to be an insurmountable technical problem. A commitment was made by both the owner and the engineer to accomplish what was a common goal. The engineer made a commitment to design and construct the wall in a manner which permitted a real-time evaluation of problems and the ability to take mitigating measures throughout the construction of the wall. The owner made a commitment to give the engineer both the professional latitude and resources required to perform his work. A design-construct contract was negotiated whereby the design could be modified as actual conditions were encountered during construction. But even with all of the planning, investigation and design efforts, there still remained a sizable risk of failure.

The wall was successfully built—not according to a pre-devised plan which went smoothly, and not without numerous problems to be resolved as unexpected groundwater and geological conditions were encountered. Estimated costs were exceeded as each unexpected condition was addressed. But there were no construction delays and their attendant costs as disputes over changed conditions and contract terms were reconciled. There were no costs for legal fees arising from litigation nor increased interest costs as construction stopped while disputes were litigated. The owner paid more than was estimated, but not more than was necessary and not as much as if he had to acquire the property at the top of the hill to regrade the slope. In addition, the owner was able to attain the desired facility expansion in far less time than by any other method.

As a result of the success of this experience and others, the use of tie-back retaining walls has become a routine practice.

Construction Site Environment

While the general information about the construction site is usually available at the planning stage of a project, it is important for the design professionals and construction manager as well as the contractor to visit the site. Each group will be benefited by first-hand knowledge acquired in the field.

For design professionals, an examination of the topography may focus their attention to the layout of a facility on the site for maximum use of space in compliance with various regulatory restrictions. In the case of industrial plants, the production or processing design and operation often dictate the site layout. A poor layout can cause construction problems such as inadequate space for staging, limited access for materials and personnel, and restrictions on the use of certain construction methods. Thus, design and construction inputs are important in the layout of a facility.

The construction manager and the contractor must visit the site to gain some insight in preparing or evaluating the bid package for the project. They can verify access roads and water, electrical and other service utilities in the immediate vicinity, with the view of finding suitable locations for erecting temporary facilities and the field office. They can also observe any interferences of existing facilities with construction and develop a plan for site security during construction.

In examining site conditions, particular attention must be paid to environmental factors such as drainage, groundwater and the possibility of floods. Of particular concern is the possible presence of hazardous waste materials from previous uses. Cleaning up or controlling hazardous wastes can be extremely expensive.

Groundwater Pollution from a Landfill

The presence of waste deposits on a potential construction site can have substantial impacts on the surrounding area. Under existing environmental regulations in the United States, the responsibility for cleaning up or otherwise controlling wastes generally resides with the owner of a facility in conjunction with any outstanding insurance coverage.

The landfill consisted of general refuse and was covered by a very sandy material. The inevitable infiltration of water from the surface or from the groundwater into the landfill will result in vertical or horizontal percolation of leachable ions and organic contamination. This leachate would be odorous and potentially hazardous in water. The pollutant would show up as seepage downhill, as pollution in surface streams, or as pollution entering the regional groundwater.

Before new construction could proceed, this landfill site would have to be controlled or removed. Typical control methods might involve:

- Surface water control measures, such as contour grading or surface sealing.
- Passive groundwater control techniques such as underground barriers between the groundwater and the landfill.
- Plume management procedures such as pumping water from surrounding wells.
- Chemical immobilization techniques such as applying surface seals or chemical injections.
- Excavation and reburial of the landfill requiring the availability of an engineered and environmentally sound landfill.

The excavation and reburial of even a small landfill site can be very expensive. For example, the estimated reburial cost for a landfill was in excess of $ 4 million in 1978.

Value Engineering

Value engineering may be broadly defined as an organized approach in identifying unnecessary costs in design and construction and in soliciting or proposing alternative design or construction technology to reduce costs without sacrificing quality or performance requirements. It usually involves the steps of gathering pertinent information, searching for creative ideas, evaluating the promising alternatives, and proposing a more cost effective alternative. This approach is usually applied at the beginning of the construction phase of the project life cycle.

The use of value engineering in the public sector of construction has been fostered by legislation and government regulation, but the approach has not been widely adopted in the private sector of construction. One explanation may lie in the difference in practice of engineering design services in the public and private sectors. In the public sector, the fee for design services is tightly monitored against the "market price," or may even be based on the lowest bid for service. Such a practice in setting professional fees encourages the design professionals to adopt known and tried designs and construction technologies without giving much thought to alternatives that are innovative but risky.

Contractors are willing to examine such alternatives when offered incentives for sharing the savings by owners. In the private sector, the owner has the freedom to offer such incentives to design professionals as well as the contractors without being concerned about the appearance of favoritism in engaging professional services.

Another source of cost savings from value engineering is the ability of contractors to take advantage of proprietary or unusual techniques and knowledge specific to the contractor's firm. For example, a contractor may have much more experience with a particular method of tunneling that is not specified in the original design and, because of this experience, the alternative method may be less expensive. In advance of a bidding competition, a design professional does not know which contractor will undertake the construction of a facility. Once a particular contractor is chosen, then modifications to the construction technology or design may take advantage of peculiar advantages of the contractor's organization.

As a final source of savings in value engineering, the contractor may offer genuine new design or construction insights which have escaped the attention of the design professional even if the latter is not restrained by the fee structure to explore more alternatives. If the expertise of the contractor can be utilized, of course, the best time to employ it is during the planning and design phase of the project life cycle. That is why professional construction management or integrated design/construction are often preferred by private owners.

Construction Planning

The development of a construction plan is very much analogous to the development of a good facility design. The planner must weigh the costs and reliability of different options while at the same time insuring technical feasibility. Construction planning is more difficult in some ways since the building process is dynamic as the site and the physical facility change over time as construction proceeds. On the other hand, construction operations tend to be fairly standard from one project to another, whereas structural or foundation details might differ considerably from one facility to another.

Forming a good construction plan is an exceptionally challenging problem. There are numerous possible plans available for any given project. While past experience is a good guide to construction planning, each project is likely to have special problems or opportunities that may require considerable ingenuity and creativity to overcome or exploit. Unfortunately, it is quite difficult to provide direct guidance concerning general procedures or strategies to form good plans in all circumstances.

There are some recommendations or issues that can be addressed to describe the characteristics of good plans, but this does not necessarily tell a planner how to discover a good plan. However, as in the design process, strategies of decomposition in which planning is divided into subproblems and hierarchical planning in which general activities are repeatably subdivided into more specific tasks can be readily adopted in many cases.

From the standpoint of construction contractors or the construction divisions of large firms, the planning process for construction projects consists of three stages that take place between the moment in which a planner starts the plan for the construction of a facility to the moment in which the evaluation of the final output of the construction process is finished.

The estimate stage involves the development of a cost and duration estimate for the construction of a facility as part of the proposal of a contractor to an owner. It is the stage in which assumptions of resource commitment to the necessary activities to build the facility are made by a planner. A careful and thorough analysis of different conditions imposed by the construction project design and by site characteristics are taken into consideration to determine the best estimate. The success of a contractor depends upon this estimate, not only to obtain a job but also to construct the facility with the highest profit. The planner has to look for the time-cost combination that will allow the contractor to be successful in his commitment.

The result of a high estimate would be to lose the job, and the result of a low estimate could be to win the job, but to lose money in the construction process. When changes are done, they should improve the estimate, taking into account not only present effects, but also future outcomes of succeeding activities. It is very seldom the case in which the output of the construction process exactly echoes the estimate offered to the owner.

In the *monitoring and control stage* of the construction process, the construction manager has to keep constant track of both activities' durations and ongoing costs. It is misleading to think that if the construction of the facility is on schedule or ahead of schedule, the cost will also be on the estimate or below the estimate, especially if several changes are made. Constant evaluation is necessary until the construction of the facility is complete. When work is finished in the construction process, and information about it is provided to the planner, the third stage of the planning process can begin.

The *evaluation* stage is the one in which results of the construction process are matched against the estimate. A planner deals with this uncertainty during the estimate stage. Only when the outcome of the construction process is known is he/she able to evaluate the validity of the estimate. It is in this last stage of the planning process that he or she determines if the assumptions were correct. If they were not or if new constraints emerge, he/she should introduce corresponding adjustments in future planning.

Industrialized Construction and Pre-fabrication

Another approach to construction innovation is to apply the principles and organizational solutions adopted for manufacturing. Industrialized construction and pre-fabrication would involve transferring a significant portion of construction operations from the construction site to more or less remote sites where individual components of buildings and structures are produced. Elements of facilities could be prefabricated off the erection site and assembled by cranes and other lifting machinery.

There are a wide variety and degrees of introducing greater industrialization to the construction process. Many components of constructed facilities have always been manufactured, such as air conditioning units. Lumber, piping and other individual components are manufactured to standard sizes. Even temporary items such as forms for concrete can be assembled off-site and transported for use. Reinforcing bars for concrete can also be pre-cut and shaped to the desired configuration in a manufacturing plant or in an automated plant located proximate to a construction site.

A major problem in extending the use of pre-fabricated units is the lack of standardization for systems and building regulations. While designers have long adopted standard sizes for individual components in designs, the adoption of standardized sub-assemblies is rarer. Without standardization, the achievement of a large market and scale economies of production in manufacturing may be impossible. An innovative and more thorough industrialization of the entire building process may be a primary source of construction cost savings in the future.

Planning of pre-fabrication

When might pre-fabricated components be used in preference to components assembled on a construction site? A straightforward answer is to use pre-fabricated components whenever their cost, including transportation, is less than the cost of assembly on site.

As an example, forms for concrete panels might be transported to a construction site with reinforcing bars already built in, necessary coatings applied to the forms, and even special features such as electrical conduit already installed in the form. In some cases, it might be less expensive to pre-fabricate and transport the entire concrete panel to a manufacturing site. In contrast, traditional construction practice would be to assemble all the different features of the panel on-site. The relevant costs of these alternatives could be assessed during construction planning to determine the lowest cost alternative.

In addition to the consideration of direct costs, a construction planner should also consider some other aspects of this technology choice.

First, the planner must insure that pre-fabricated components will satisfy the *relevant building codes* and *regulations*. Second, the *relative quality* of traditional versus pre-fabricated components as experienced in the final facility should be considered. Finally, the *availability of components* at the required time during the construction process should also be considered.

Impacts of building codes

Building codes originated as a part of the building regulatory process for the safety and general welfare of the public. The source of all authority to enact building codes is based on the police power of the state which may be delegated by the state legislature to local government units.

Consequently, about 8,000 localities having their own building codes, either by following a national model code or developing a local code. The lack of uniformity of building codes may be attributed to a variety of reasons:

- Neighboring municipalities may adopt different national models as the basis for local regulation.
- Periodic revisions of national codes may not be adopted by local authorities before the lapse of several years.
- Municipalities may explicitly decline to adopt specific provisions of national model codes or may use their own variants of key provisions.
- Local authorities may differ in interpretation of the same language in national model codes.

The lack of uniformity in building codes has serious impact on design and construction as well as the regulatory process for buildings. Among the significant factors are:

- Delay in the diffusion of new building innovations which may take a long time to find their ways to be incorporated in building codes.
- Discouragement to new production organizations, such as industrialized construction and prefabrication.
- Duplication of administrative cost of public agencies and compliance cost incurred by private firms.

Computer-Aided Engineering

In the past twenty years, the computer has become an essential tool in engineering, design, and accounting. The innovative designs of complicated facilities cited in the previous sections would be impossible without the aid of computer based analysis tools. By using general purpose analysis programs to test alternative designs of complex structures such as petrochemical plants, engineers are able to greatly improve initial designs. General purpose accounting systems are also available and adopted in organizations to perform routine bookkeeping and financial accounting chores. These applications exploit the capability for computers to perform numerical calculations in a pre-programmed fashion rapidly, inexpensively and accurately.

Despite these advances, the computer is often used as only an incidental tool in the design, construction and project management processes. However, new capabilities, systems and application programs are rapidly being adopted. These are motivated in part by the remarkable improvement in computer hardware capability, the introduction of the Internet, and an extraordinary decline in cost.

New concepts in computer design and in software are also contributing. For example, the introduction of personal computers using microcircuitry has encouraged the adoption of interactive programs because of the low cost and considerable capability of the computer hardware. Personal computers available for a thousand dollars in 1995 have essentially the same capability as expensive mainframe computer systems of fifteen years earlier.

Computer graphics provide another pertinent example of a potentially revolutionary mechanism for design and communication. Graphical representations of both the physical and work activities on projects have been essential tools in the construction industry for decades. However, manual drafting of blueprints, plans and other diagrams is laborious and expensive. Stand alone, computer aided drafting equipment has proved to be less expensive and fully capable of producing the requiring drawings.

More significantly, the geometric information required for producing desired drawings might also be used as a database for computer aided design and computer integrated construction. Components of facilities can be represented as three dimensional computer based *solid models* for this purpose. Geometric information forms only one component of integrated design databases in which the computer can assure consistency, completeness and compliance with relevant specifications and constraints. Several approaches to integrated computer aided engineering environments of this type have already been attempted.

Computers are also being applied more and more extensively to non-analytical and non-numerical tasks. For example, computer based specification writing assistants are used to rapidly assemble sets of standard specifications or to insert special clauses in the documentation of facility designs.

As another example, computerized transfer of information provides a means to avoid laborious and error-prone transcription of project information. While most of the traditional applications and research in computer aids have emphasized numerical calculations, the use of computers will rapidly shift towards the more prevalent and difficult problems of planning, communication, design and management.

Knowledge based systems represent a prominent example of new software approaches applicable to project management. These systems originally emerged from research in artificial intelligence in which human cognitive processes were modeled. In limited problem domains such as equipment configuration or process control, knowledge based systems have been demonstrated to approach or surpass the performance of human experts.

The programs are marked by a separation between the reasoning or "inference" engine program and the representation of domain specific knowledge. As a result, system developers need not specify complete problem solving strategies (or algorithms) for particular problems. This characteristic of knowledge based systems make them particularly useful in the ill-structured domains of design and project management.

Computer program assistants will soon become ubiquitous in virtually all project management organizations. The challenge for managers is to use the new tools in an effective fashion. Computer intensive work environments should be structured to aid and to amplify the capabilities of managers rather than to divert attention from real problems such as worker motivation.

Pre-Project Planning

Even before design and construction processes begin, there is a stage of "pre-project planning" that can be critical for project success. In this process, the project scope is established. Since construction and design professionals are often not involved in this project scope stage, the terminology of describing this as a "pre-project" process has arisen. From the owner's perspective, defining the project scope is just another phase in the process of acquiring a constructed facility.

The definition of a project scope typically involves developing project alternatives at a conceptual level, analyzing project risks and economic payoff, developing a financial plan, making a decision to proceed (or not), and deciding upon the project organization and control plan. The next few chapters will examine these different problems at some length.

The danger of poor project definition comes from escalating costs (as new items are added) or, in the extreme, project failure. A good definition of scope allows all the parties in the project to understand what is needed and to work towards meeting those needs.

GLOSSARY

Abutment: Supporting structure at each end of a single span bridge. Types of abutments vary based on geotechnical, geographic, and constructability factors.

Actuator: A device operated either electrically, pneumatically, or hydraulically that acts as a motor to change the position of movable devices such as valves or dampers.

Addendum: Any documentation (drawings, specifications, etc.) issued during the bidding process to modify, clarify, or complement the information outlined in the bidding documents. Addenda become part of the contract documents upon awarding of the contract to a firm.

As-builts: The common term for marked up floor plans (usually in red ink) showing how installation differed from the original design drawings. As-builts are usually created for utilities and in-wall systems that aren't directly visible. The formal term for as-builts is record drawings.

Back charge: A charge against a contractor's contract for costs incurred by another party that should have been incurred by the contractor. A back charge typically takes form in a deductive change order. For example, if the contractor in charge of fire sprinklers damaged a wall using a boom lift, that contractor may incur a back charge from the drywall contractor for the repair work.

Backfilling: This is to refill excavated trenches or holes with the material dug out of it.

Beam: A sturdy piece of timber or metal, usually horizontal, with a main function to support loads such as the roof or floor above.

Bid: A binding offer made by a contractor to do the scope of work specified in the bidding documents at a certain price. This offer must be in accordance with the plans and specifications of the project and the terms and conditions stated in the offer.

Building automation system (BAS): An energy management system, usually with additional capabilities, relating to the overall operation of the building in which it is installed; this includes equipment monitoring, protection of equipment against power failure, and building security. It's the digital automation that takes operation of an HVAC design out of someone's physical hands and moves it onto a computer interface.

Building management system (BMS): An energy management system relating to the overall operation of the building in which it is installed. It often has additional capabilities, such as equipment monitoring, protection of equipment against power failure, and building security. It may also be a direct digital control (DDC) system where the mode of control uses digital outputs to control processes or elements directly.

Bulletins: The new drawings or specifications that are issued (usually by the architect or lead designer) to the contractor after its contract award. These may be a result of RFIs, an owner's request, errors/omissions, or a reviewing agency request.

Cantilever: A beam which is securely supported at one end, and hangs freely at the other, like balconies that have no pillars or columns supporting them.

Change order (CO): A change to the original contract scope of work, resulting in either an increase or decrease in project costs and/or schedule. This could be issued by a client to the general contractor or the general contractor to a subcontractor.

Clear and grub: The process of removing vegetation and material from land in the construction area. This prepares the site for excavation and grading work to take place.

Close out: A phase that encompasses the end of a project. This consists of punch lists, O&M manuals, as-builts/record drawings, warranties, and the overall completion of project work.

Cofferdam: A structural enclosure installed to retain water in damp soil or bodies of water. The cofferdam is installed and water is pumped out of the enclosure to provide a dry and safe work environment.

Commissioning: The process of testing and verifying the intended behavior of the building systems, such as HVAC, plumbing, electrical, life safety, lighting, etc. This is generally conducted by the contractor in tandem with the facility management team in order to ensure that the building staff are prepared to operate and maintain its systems and equipment.

Computer-Based System: A management system in which a computer is the central controlling device.

Construction daily reports (CDR): A record of field notes, including work completed, weather conditions, and materials delivered or used on-site.

Contingencies: The cost provisions in the project budget that make allowance for oversights and unforeseen circumstances associated with the project. Depending on the nature of the contract, the contractor may require owner's approval to draw funds from contingencies.

Contract drawings: The conformed set of drawings that represents the entire contract scope.

Core & shell: A term that refers to the base construction of a building. This includes its structure, cladding, and vertical systems, such as MEP utility risers, stairs, and elevators, as well as its finished common areas, such as lobby and restrooms.

Critical lift: Per OSHA 1926.751, a lift that (1) exceeds %75 of the rated capacity of the crane or derrick, or (2) requires the use of more than one crane or derrick.

Damp-proof course: Usually abbreviated as DPC, this is a watertight layer applied to masonry to prevent moisture rising up from the ground.

Datum: Also known as datum point, this refers to any elevation taken as a reference point for leveling.

Derrick barge: Also known as a crane barge. These floating cranes lift large loads while on water. These are great for offshore construction use.

Direct digital control (DDC): A type of control where controlled and monitored analog or binary data (e.g., temperature, contact closures) are converted to digital format for manipulation and calculations by a digital computer or micro-processor, then converted back to analog or binary form to control physical devices.

DLB (dig, lay, bury/backfill): Term used in underground construction that describes the activity of utility installation. Dig/excavate dirt, lay your utility line, and backfill.

Embankment: An elevated landmass made of compacted soil or aggregate. Embankments help create roads and level surfaces for slabs or levees.

Energy management system (EMS): A system of computer applications used by building engineering staff to monitor, control, and optimize the building's operating performance (e.g., energy consumption, occupant comfort levels). EMS optimizes building operating performance through supervisory control programs that utilize core BMS functionality. An EMS is a more refined definition of a building management system (BMS) focused exclusively on operating performance with things like indoor air quality (IAQ), occupant productivity and comfort, and energy output as key metrics.

Falsework: These are the temporary structures that are built to hold and support cast-in-place concrete during pours and while it cures to strength.

Field order: A work order issued to a contractor by the owner or general contractor to perform work not included in the contract. The contractor will then be entitled to a change order for the extra work. Field orders are used to expedite work in an emergency or crucial situation, when there is no time to compile and approve a change order request.

Fit out: A term that refers to the interior construction of a building to make it suitable for occupation. This could include distribution of MEP services, ceiling systems, finishes, furniture, lighting, etc.

Grade and compact: The process of leveling out an area of land and exerting force using heavy equipment to stabilize the soil or aggregate.

Hip roof: A roof where all four sides slope down towards the walls.

HVAC: The equipment, terminals, and distribution systems that provide, either collectively or individually, the processes of heating, ventilating, or air conditioning to a building or portion of a building. Basically, the setup of machines that make a room or space warmer, cooler, or more conditioned in some regard.

HVAC Zone: A space or group of spaces within a building with heating, cooling, and ventilating requirements that are sufficiently similar so that desired conditions (e.g., temperature) can be maintained throughout using a single sensor (e.g., thermostat or temperature sensor).

IR: This stands for "inspection request" which is a form required by independent third-party inspectors to confirm an installation detail or method. It is often used for work such as welding, anchoring, and concrete pours.

Kip: A unit of force or weight, equal to 1,000 pounds, used to measure engineering loads.

K-rail/Jersey barrier: A precast concrete barrier that comes in 10' or 20' lengths. These are used to separate traffic, protect traffic from a leading edge, or designate the construction zone.

Liquidated damages: An amount of money that the contractor would owe the owner in the event of a breach of contract. These are typically calculated by a formula such as 10,000$ per day of substantial delay of completion.

MSE (mechanically stabilized earth) wall: A wall system consisting of vertical panels and horizontal straps backfilled with aggregate or soil. MSE walls allow for easier and quicker installation than traditional cast-in-place methods.

OAC meeting: A meeting held at a scheduled time (generally weekly or biweekly) between the owner, the architect, and the general contractor. This meeting covers general project management topics, such as safety, scheduling, procurement, RFIs, submittals, change orders, design changes, etc. The general contractor is usually responsible for leading the meeting and distributing the meeting minutes.

O&M manual: This stands for operations and maintenance manual which is an instruction guide to running and servicing a building. The O&M manual is given to the owner at turnover.

Pile: Usually, a long structural member driven into the ground to act as support for a foundation or wall. There are many forms of pile that differ by material, shape, and size.

PPE: This stands for "personal protective equipment" which is basically the proper attire for a construction job. Such equipment often includes a hard hat, hard-soled boots, reflective vest, safety glasses, long pants, and a shirt with at least 6» long sleeves.

Post-tensioning: A method for prestressing (strengthening) concrete whereby cables are pulled or the concrete is jacked up after it has been placed.

Preliminary notice: Documents notifying other parties that they are working on the project. They're often required to preserve mechanics lien rights. But even when not required, they provide invaluable transparency and communication.

Procurement log: A spreadsheet that tracks how long it will take for certain materials to be delivered on-site, especially those with long lead times. It tracks when the submittals for these materials need to be approved so that they can be ordered for fabrication and delivery. Some examples of items that may be on a procurement log are structural steel, HVAC equipment, lighting fixtures, and custom cabinetry.

Project schedule: The contract timeline for the duration of the project, usually in Gantt chart format showing relationships and the critical path (what items are driving the overall schedule). It is generally composed of long duration tasks, sorted by scope and location, as opposed to the 3 week look ahead (which is very detailed and short term).

Punch list: A list of all items that need to be fixed before the building or project can be turned over to the client. Punch list items are also known as snags or deficiencies, and include things like paint scratches, damaged siding, cleanup, etc. This process comes at the end of the project after a preliminary walkthrough of the jobsite. The final punch list is usually tied to a cost withheld from the contractor until it is completed and verified.

QA/QC: This stands for "quality assurance/quality control" which is the formalized process of confirming proper installation methods and materials on-site.

RFI: This stands for "request for information" which is a formal question to the design team, client, or general contractor that will likely affect the contract scope, drawings, and/or specifications. The response to an RFI acts as a contract change and could result in a change order.

Rough-in: The initial stage of the wall framing, HVAC, electrical, and plumbing installation. This includes all of the components that won't be seen after the completion of the project. All trade rough-ins must generally be inspected prior to insulation and application of finishes.

Schedule of values: The breakdown of a contract amount into sub-items and sub-costs for identifiable construction elements. This is usually used as the basis for submitting and reviewing progress payment.

Sensor: A device or instrument designed to detect and measure a variable.

Shoring: Temporary structural support for underground excavation to prevent soil movement or collapse. Shoring is used when you can no longer excavate to your maximum allowable slope.

Shotcrete: Concrete that is shot onto a reinforced surface, usually wire mesh, by way of air pressure and hose. Shotcrete is a typically installed on soil nail walls.

Specifications: The instruction manual that is paired with the contract drawings. It includes information such as installation procedures, product brand requirements, and testing requirements, to name a few. It is separated out by CSI (Construction Specification Institute) divisions, generally starting with 00 00 01 through 00 00 33.

Strut: A structural member, usually inclined and shorter in length, used to resist axial force.

Submittal: A document that is prepared by the installing contractor that indicates the products and locations that will be procured and installed. The document will need to be approved by the architect (and other applicable design team members), as well as the general contractor and client. An example of a submittal would be a cut sheet of the sink that will be installed in the bathrooms.

Substantial completion: A milestone in construction projects defined as the stage when work is sufficiently completed in accordance with the contract documents. This indicates that the owner can now utilize the building or facility for its intended purpose. Only minor works, such as punch list items, will ordinarily remain after reaching substantial completion.

Take off: An estimation of the quantity of material required to complete a certain scope of work.

Toothing: When alternating bricks are left projecting from walls in order to bond with future work.

Tremie concrete: Concrete that is designed to be poured underwater through a gravity feed system. This method is often used to create underwater foundations as well as seal cofferdams and caissons.

Variable-frequency drive (VFD): An electronic device that varies its output frequency to vary the rotating speed of a motor given a fixed input frequency. They are used with fans or pumps to vary the flow in the system as a function of a maintained pressure.

BIBLIOGRAPHY

Adkins, L.; Cooper, M.; Konings, M. Class in the 21st century: Asset inflation and the new logic of inequality. Environ. Plan. A Econ. Space 2019.

Alam, T. A reliable communication framework and its use in internet of things (IOT). Int. J. Sci. Res. Comput. Sci. Eng. Inf. Techcnol. 2018, 3, 450–456.

Alexander, C. The Nature of Order: The Process of Creating Life; The Centre for Environmental Structure: Berkeley, CA, USA, 2002.

Alexander, C.; Ishikawa, S.; Silverstein, M. A Pattern Language; Oxford University Press: New York, NY, USA, 1977.

Allam, Z. Contextualising the smart city for sustainability and inclusivity. New Des. Ideas 2018, 2, 124–127.

Allam, Z. Surveying the Covid-19 Pandemic and Its Implications: Urban Health, Data Technology and Political Economy; Elsevier Science: Amsterdam, The Netherlands, 2020.

Allam, Z.; Dhunny, Z.A. On big data, artificial intelligence and smart cities. Cities 2019, 89, 80–91.

Allam, Z.; Jones, D.; Thondoo, M. Cities and Climate Change: Climate Policy, Economic Resilience and Urban Sustainability; Springer International Publishing: Cham, Switzerland, 2020.

Allam, Z.; Jones, D.S. Pandemic stricken cities on lockdown. Where are our planning and design professionals [now, then and into the future]? Land Use Policy 2020, 97, 104805. [PubMed]

Allam, Z.; Newman, P. Redefining the smart city: Culture, metabolism and governance. Smart Cities 2018, 1, 4–25.

Allen, C.; Clouth, S. A Guidebook to the Green Economy; UNDESA: New York, NY, USA, 2012.

Appleyard, D. Livable streets: Protected neighborhoods? Ann. Am. Acad. Political Soc. Sci. 1980, 451, 106–117.

Beatley, T. Biophilic Cities: Integrating Nature into Urban Design and Planning; Island Press: Washington, DC, USA, 2010.

Beatley, T. Green Urbanism: Learning from European Cities; Island Press: Washington, DC, USA, 2000.

Beck, U. Risk Society. Towards a New Modernity; Sage: London, UK, 2000.

Bioregional. The BedZED Story. The UK's First Large Scale, Mixed-Use Eco-Village; Bioregional Development Group: London, UK, 2016.

Bolt, W.; Van Oordt, M.R.C. On the value of virtual currencies. J. Money Credit Bank. 2020, 52, 835–862.

Boselli, F. Cities Must Be Regenerative. But What Kind of Regeneration Are We Actually Talking About? World Future Council. 2016. Available online: https://www.worldfuturecouncil.org/cities-must-regenerative-kind-regeneration-actually-talking/ (accessed on 12 April 2017).

Bouton, S.; Hannon, E.; Knupfer, S.; Ramkumar, S. The Future (s) of Mobility: How Cities Can Benefit. Available online: https://www.mckinsey.com/business-functions/sustainability/our-insights/the-futures-of-mobility-how-cities-can-benefit (accessed on 15 January 2020).

Brandon, E.M. 7 Smart and Simple Ways Cities Can Encourage Social Distancing in Public Spaces. Available online: https://citymonitor.ai/community/smart-and-simple-ways-public-spaces-are-being-adapted-for-social-distancing (accessed on 12 November 2020).

Brookfield, K. Residents' preferences for walkable neighbourhoods. J. Urban Des. 2017, 22, 44–58.

Brown, J.R.; Morris, E.A.; Taylor, B.D. Planning for cars in cities: Planners, engineers, and freeways in the 20th century. J. Am. Plan. Assoc. 2009, 75, 161–177.

Brown, L.R. Redesigning cities for people: Car-centered urban sprawl. In Eco-Economy: Building an Economy for the Earth; Earthscan: Washington, DC, USA, 2003.

Bulkeley, H.; Broto, V.C.; Hodson, M.; Marvin, S. (Eds.) Cities and Low Carbon Transitions; Routledge: London, UK, 2010.

C40 Cities Climate Leadership Group. How to Build Back Better with a 15-Minute City.

Campbell, H. How Paris is Actually Walking the Climate Change Walk. Available online: https://time.com/5669067/paris-green-city/ (accessed on 10 November 2020).

Capasso Da Silva, D.; King, D.A.; Lemar, S. Accessibility in practice: 20-minute city as a sustainability planning goal. Sustainability 2020, 12, 129.

Capra, F.; Spretnak, C. Green Politics: The Global Promise; Dutton: New York, NY, USA, 1984.

Caragliu, A.; Del Bo, C.; Nijkamp, P. Smart Cities in Europe. J. Urban Technol. 2011, 18, 65–82.

Causone, F.; Sangalli, A.; Pagliano, L.; Carlucci, S. Assessing energy performance of smart cities. Build. Serv. Eng. Res. Technol. 2017, 39, 99–116.

Cervero, R. Efficient urbanisation: Economic performance and the shape of the metropolis. Urban Stud. 2001, 38, 1651–1671.

Cervero, R.; Kockelman, K. Travel demand and the 3ds: Density, diversity, and design. Transp. Res. Part D Transp. Environ. 1997, 2, 199–219.

Chel, A.; Kaushik, G. Renewable energy technologies for sustainable development of energy efficient building. Alex. Eng. J. 2018, 57, 655–669.

City of Vancouver. City Repurposes Eastbound Lanes of Beach Ave for People Walking and Cycling. Available online: https://vancouver.ca/news-calendar/city-repurposes-eastbound-lanes-of-beach-ave-for-people-walking-and-cycling.aspx (accessed on 5 November 2020).

Cocchia, A. Smart and digital city: A systematic literature review. In Smart City; Springer International Publishing: Cham, Switzerland, 2014; pp. 13–43.

Cohen, J.; Stewart, I. The Collapse of Chaos: Discovering Simplicity in a Complex World; Penguin Group Ltd.: London, UK, 1994.

Cohen, J.-L.; Benton, T. (Eds.) Le Corbusier Le Grand; Phaidon Press Limited: London, UK, 2008.

Cohen, N.; Rubinchik, A.; Shami, L. Towards a cashless economy: Economic and socio-political implications. Eur. J. Political Econ. 2020, 61, 101820.

Commonwealth of Australia. Tackling Wicked Problems; a Public Policy Perspective; Australian government/Australian Public Service Commission: Canberra, Australia, 2007.

Connolly, K. 'Cleaner and Greener': Covid-19 Prompts World's Cities to Free Public Space of Cars. Available online: https://www.theguardian. com/world/2020/may/18/cleaner-and-greener-covid-19-prompts-worlds-cities-to-free-public-space-of-cars (accessed on 5 November 2020).

Corbin, J.; Strauss, A. Basics of Qualitative Research; Sage Publications: Thousand Oaks, CA, USA, 2008.

Coulon, J. Oakland is Closing Streets to Most Cars. More Cities Need to Do the Same. Available online: https://www.bicycling.com/news/a32130879/oakland-temporary-street-closures-coronavirus/ (accessed on 5 November 2020).

Crang, M. Rhythms of the city: Temporalised space and motion. In Timespace: Geographies of Temporality; May, J., Thrift, N., Eds.; Routledge: London, UK, 2001; pp. 187–207.

Davies, S. This is How Coronavirus Could Reshape Our Cities Forever. Available online: https://www.weforum.org/agenda/2020/05/coronavirus-change-cities-infrastructure/ (accessed on 4 November 2020).

De Jong, M.; Joss, S.; Schraven, D.; Zhan, C.; Weinen, M. Sustainable-smart-resilient-low carbon-eco-knowledge cities; making sense of a multitude of concepts promoting sustainable urbanization. J. Clean. Prod. 2015, 109, 25–38.

De Vos, J. The effect of Covid-19 and subsequent social distancing on travel behavior. Transp. Res. Interdiscip. Perspect. 2020, 5, 100121.

DeLisle, J.R.; Grissom, T.V. An empirical study of the efficacy of mixed-use development: The Seattle experience. J. Real Estate Lit. 2013, 21, 25–57.

Dembski, F.; Wössner, U.; Letzgus, M.; Ruddat, M.; Yamu, C. Urban digital twins for smart cities and citizens: The case study of Herrenberg, Germany. Sustainability 2020, 12, 2307.

Dempsey, N.; Brown, C.; Bramley, G. The key to sustainable urban development in UK cities? The influence of density on social sustainability. Prog. Plan. 2012, 77, 89–141.

District Office Berlin. Temporary Installation and Expansion of Bicycle Traffic Facilities during the Pandemic Crisis. Available online: https://www.berlin.de/ba-friedrichshain-kreuzberg/aktuelles/pressemit teilungen/2020/pressemitteilung.911780.php (accessed on 5 November 2020).

Donatiello, J.E. The world's population: An encyclopedia of critical issues, crises, and ever-growing countries. Ref. User Serv. Q. 2015, 54, 85.

Duany, A.; Plater-Zyberk, E.; Speck, J. Suburban Nation: The Rise of Sprawl and the Decline of the American; Dream North Point: New York, NY, USA, 2000.

Duany, A.; Speck, J.; Lydon, M. The Smart Growth Manual; McGraw-Hill Education: New York, NY, USA, 2009.

Ecobuilders (Undated). What Is an Ecocity? Available online: http://www.ecocitybuidlers.org/what-is-an-ecocity/ (accessed on 12 April 2017).

Ellis, C. Landscape Urbanism and New Urbanism: A View of the Debate. J. Urban Des. 2015, 20, 303–307.

Ewing, R.; Cervero, R. Travel and the built environment. J. Am. Plan. Assoc. 2010, 76, 265–294.

Farr, D. Sustainable Urbanism. Urban Design with Nature; John Wiley and Sons Inc.: Hoboken, NJ, USA, 2008.

Ferrero Bermejo, J.; Gómez Fernández, J.F.; Olivencia Polo, F.; Crespo Márquez, A. A review of the use of artificial neural network models for energy and reliability prediction. A study of the solar pv, hydraulic and wind energy sources. Appl. Sci. 2019, 9, 1844.

Folch-Serra, M. Place, voice, space: Mikhail Bakhtin's dialogical landscape. Environ. Plan. D Soc. Space 1990, 8, 255–274.

Foxon, T.J.; Hammond, G.P.; Pearson, P.J. Developing transition pathways for a low carbon electricity system in the UK. Technol. Forecast. Soc. Chang. 2010, 77, 1203–1213.

Gehl, J. Cities for People; Island Press: Washington, DC, USA, 2013.

Girardet, H. The GAIA Atlas of Cities: New Directions for Sustainable Urban Living; Gaia Books Limited: London, UK, 1996.

Glaser, B.; Strauss, A. The Discovery of Grounded Theory; Aldine: Chicago, IL, USA, 1967.

Gössling, S. Why cities need to take road space from cars—And how this could be done. J. Urban Des. 2020, 25, 443–448.

Gunder, M. Fake it until you make it, and then.... Plan Theory 2011, 10, 201–212.

Gurstein, P.; Hutton, T. Planning on the Edge: Vancouver and the Challenges of Reconciliation, Social Justice, and Sustainable Development; UBC Press: Vancouver, BC, Canada, 2019.

Haas, T. Sustainable Urbanism and Beyond; Rizzoli: New York, NY, USA, 2012.

Hammad, A.W.; Akbarnezhad, A.; Haddad, A.; Vazquez, E.G. Sustainable zoning, land-use allocation and facility location optimisation in smart cities. Energies 2019, 12, 1318.

Han, C.; Wu, Y.; Chen, Z.; Wang, X. Terahertz communications (TeraCom): Challenges and impact on 6G wireless systems. arXiv 2019, arXiv:1912.06040.

Haughton, G. Developing sustainable urban development models. Cities 1997, 14, 189–195.

Herrera, S.; Pasztor, A. Amazon gets U.S. Approval for Drone Fleet, a Package-Delivery Milestone. Available online: https://www.wsj.com/articles/amazon-gets-u-s-approval-for-drone-fleet-a-package-delivery-milestone-11598913514 (accessed on 12 November 2020).

Howard, E. Garden Cities of Tomorrow; S. Sonnenschein & Co., Ltd.: London, UK, 1902.

Howarth, L. Ecopolis, the Emergence of Regenerative Cities. The Ecologist. Available online: http://www.theecologist.org/News/news_analysis/2000416/ecopolis_the_emergence_of_regenerative_cities.html (accessed on 12 April 2017).

Hu, W. A Surge in Biking to Avoid Crowded Trains in N.Y.C. Available online: https://www.nytimes.com/2020/03/14/nyregion/coronavirus-nyc-bike-commute.html (accessed on 5 November 2020).

Ibold, S.; Medimorec, N.; Wagner, A.; Peruzzo, J.; Platzer, L. The Covid-19 Outbreak and Implications to Sustainable Urban Mobility—Some Observations. Available online: https://www.transformative-mobility.org/news/the-covid-19-outbreak-and-implications-to-public-transport-some-observations (accessed on 5 November 2020).

Intergovernmental Panel on Climate Change (IPCC). Climate Change 2014: Synthesis Report; Contribution of Working Groups I, II and III to the Fifth Assessment Report of the Intergovernmental Panel on Climate Change; Core Writing Team, Pachauri, R.K., Meyer, L.A., Eds.; IPCC: Geneva, Switzerland, 2014; p. 151.

Jabareen, Y. Planning the resilient city: Concepts and strategies for coping with climate change and environmental risk. Cities 2013, 31, 220–229.

Jacobs, J. The Death and Life of Great American Cities; Random House: New York, NY, USA, 1961.

Joss, S. Eco-cities: The mainstreaming of urban sustainability—Key characteristics and driving forces. Int. J. Sustain. Dev. Plan. 2011, 6, 268–285.

Joss, S.; Cowley, R.; Tomozeiu, D. Towards the 'ubiquitous eco-city': An analysis of the internationalisation of eco-city policy and practice. Urban Res. Pract. 2013, 6, 54–74.

Kahn, M.E. Green Cities Urban Growth and the Environment; Brookings: Washington, DC, USA, 2006.

Kamel Boulos, M.N.; Tsouros, A.D.; Holopainen, A. 'Social, innovative and smart cities are happy and resilient': Insights from the who euro 2014 international healthy cities conference. Int. J. Health Geogr. 2015, 14, 3.

Kauffman, S. At Home in the Universe; Oxford University Press: Oxford, UK, 1995.

Kauffman, S.A. The Origin of Order: Self-Organisation and Selection in Evolution; Oxford University Press: New York, NY, USA, 1993.

Kennedy, C.; Cuddihy, J.; Engel-Yan, J. The changing metabolism of cities. J. Ind. Ecol. 2007, 11, 43–59.

Kennedy, C.; Pincetl, S.; Bunje, P. The study of urban metabolism and its applications to urban planning and design. Environ. Pollut. 2010, 159, 1965–1973. [PubMed]

Kenworthy, J.R. The eco-city: ten key transport and planning dimensions for sustainable city development. Environ. Urban. 2006, 18, 67–85.

Kourtit, K. Super-proximity and spatial development. J. Reg. Res. 2016, 36, 215–231.

Kreimer, A.; Arnold, M.; Carlin, A. (Eds.) Building Safer Cities: The Future of Disaster Risk; The International Bank for Reconstruction and Development, The World Bank: Washington, DC, USA, 2003.

Kriens, I.; Klaasen, I.T.; Rooij, R.M. De Kern van Ruimtelijke Planning and Strategie. Ruimte en Maatschappij; Technische Universiteit Delft: Delft, The Netherlands, 2008.

Kristinsson, J. Integrated Sustainable Design; Van den Dobbelsteen, A., Ed.; Delft Digital Press: Delft, The Netherlands, 2012.

Larco, N. Sustainable urban design—A (draft) framework. J. Urban Des. 2016, 21, 1–29.

Lee, S. The Case for Building Tiny House Villages during the Pandemic. Available online: https://www.realchangenews.org/2020/06/10/case-building-tiny-house-villages-during-pandemic (accessed on 5 November 2020).

Lehmann, S. Advocacy for the compact, mixed-use and walkable city: Designing smart and climate resilient places. Int. J. Environ. Sustain. 2016, 5, 1–11.

Lehmann, S. Green Urbanism: Formulating a Series of Holistic Principles. Sapiens 2010, 3, 1–17.

Lehmann, S. Sustainable urbanism: Towards a framework for quality and optimal density? Future Cities Environ. 2016, 2, 8.

Lehmann, S. The Principles of Green Urbanism. Transforming the City for Sustainability; Earthscan: London, UK, 2010.

Leichenko, R. Climate change and urban resilience. Curr. Opin. Environ. Sustain. 2011, 3, 164–168.

Liu, J. Commuters in This City Spend 119 Hours a Year Stuck in Traffic. Available online: https://www.cnbc.com/2019/09/04/commuters-in-this-city-spend-119-hours-a-year-stuck-in-traffic.html (accessed on 10 November 2020).

Looi, M.-K. Covid-19: Is a second wave hitting Europe? BMJ 2020, 371, m4113.

Maragakis, L.L. First and Second Waves of Coronavirus. Available online: https://www.hopkinsmedicine.org/health/conditions-and-diseases/coronavirus/first-and-second-waves-of-coronavirus (accessed on 5 June 2020).

Marquet, O.; Miralles-Guasch, C. The walkable city and the importance of the proximity environments for Barcelona's everyday mobility. Cities 2015, 42, 258–266.

McHarg, I.L. Design with Nature; Natural History Press: New York, NY, USA, 1969.

McKinsey & Company. Smart City Solutions: What Drives Citizen Adoption around the Globe? Mckinsey Center for Government: Singapore, 2018; pp. 1–60.

Ministerie van Landbouw Natuurbeheer en Visserij. Nota Landschap; Ministerie van LNV: Den Haag, The Netherlands, 1992.

Minneapolis Park and Recreation Board. Minneapolis Park and Recreation Board Announces Riverfront Parkway and Road Closures to Help Trail Users Maintain Social Distancing. Available online: https://www.minneapolisparks.org/news/2020/03/26/minneapolis-park-and-recreation-board-announces-riverfront-parkway-and-road-closures-to-help-trail-users-maintain-social-distancing/ (accessed on 5 November 2020).

Mitchell Waldrop, M. Complexity. The Emerging Science at the Edge of Order and Chaos; Simon and Schuster Paperbacks: New York, NY, USA, 1992.

Moreno, C. La Ville du Quart D'heure: Pour un Nouveau Chrono-Urbanisme. Available online: https://www.latribune.fr/regions/smart-cities/la-tribune-de-carlos-moreno/la-ville-du-quart-d-heure-pour-un-nouveau-chrono-urbanisme-604358.html (accessed on 3 December 2020).

Moreno, C. Droit de Cité; Humensis: Paris, France, 2020.

Morgan, M.H. Vitruvius. The Ten Books on Architecture; Harvard University Press: Cambridge, MA, USA; Oxford University Press: London, UK, 2001.

Morris, D.Z. After Coronavirus, Bicycles Will Have a New Place in City Life. Available online: https://fortune.com/2020/06/15/bicycles-coronavirus-cities-lime-citi-bike/ (accessed on 5 November 2020).

Mostafavi, M. Why Ecological Urbanism? Why Now? In Ecological Urbanism; Mostafvai, M., Doherty, G., Eds.; Lars Müller Publishers: Baden, Switzerland, 2010.

Mulíèek, O.; Osman, R.; Seidenglanz, D. Urban rhythms: A chronotopic approach to urban timespace. Time Soc. 2014, 24, 304–325.

Nabil, N.A.; Eldayem, G.E.A. Influence of mixed land-use on realizing the social capital. HBRC J. 2015, 11, 285–298.

NCTA. The Future of Super Fast Internet Is 10G. Available online: https://www.ncta.com/positions/the-future-of-super-fast-internet (accessed on 12 November 2020).

Net Information. 1G Vs. 2G Vs. 3G Vs. 4G Vs. 5G. Available online: http://net-informations.com/q/diff/generations.html (accessed on 22 July 2020).

Neuhaus, F. Urban rhythms. In Emergent Spatio-Temporal Dimensions of the City: Habitus and Urban Rhythms; Springer International Publishing: Cham, Switzerland, 2015; pp. 1–11.

Neves, A.; Brand, C. Assessing the potential for carbon emissions savings from replacing short car trips with walking and cycling using a mixed GPS-travel diary approach. Transp. Res. Part A Policy Pract. 2019, 123, 130–146.

Newman, P.; Beatley, T.; Boyer, H. Resilient Cities: Overcoming Fossil Fuel Dependence, 2nd ed.; Island Press: Washington, DC, USA, 2017.

Newman, P.; Beatley, T.; Boyer, T. Resilient Cities; Island Press: Washington, DC, USA, 2009.

Newman, P.; Kenworthy, J. The rise and fall of automobile dependence. In The End of Automobile Dependence; Island Press/Center for Resource Economics: Washington, DC, USA, 2015; pp. 1–31.

Newman, P.W.G. Sustainability and cities: Extending the metabolism model. Landsc. Urban Plan. 1999, 44, 219–226.

Nicholls, R.J. Coastal megacities and climate change. GeoJournal 1995, 37, 369–379.

OECD. OECD Policy Responses to Coronavirus (Covid-19): Cities Policy Responses; OECD: Paris, France, 2020.

Pacheco Rocha, N.; Dias, A.; Santinha, G.; Rodrigues, M.; Queirós, A.; Rodrigues, C. Smart cities and healthcare: A systematic review. Technologies 2019, 7, 58.

Pandey, N. Smart Cities Could Result in Social Inequality, Say Experts. Available online: https://www.thehindubusinessline.com/economy/smart-cities-could-result-in-social-inequality-say-experts/article9111629.ece (accessed on 11 November 2020).

Parente, M. Designing the city identity: Strategic and product design for new experiential ways of living, enabling, and interacting with the urban context. BMI 2015, 10, 62–71.

Portugali, J. Complexity theory as a link between space and place. Environ. Plan. A 2006, 38, 647–664.

Portugali, J. Self-Organisation and the City; Springer: Berlin, Germany, 2000.

Priemus, H. The network approach: Dutch spatial planning between substratum and infrastructure networks. Eur. Plan. Stud. 2007, 15, 667–686.

Prigogine, Y.; Stengers, I. Order out of Chaos. Man's New Dialogue with Nature; Bantam Books Inc.: New York, NY, USA, 1984.

Rapoport, E.; Vernay, A.-L. Defining the Eco-City: A Discursive Approach; International Ecocities Initiative; University of Westminster: London, UK, 2014; Available online: https://www.westminster.ac.uk/file/42096/download (accessed on 15 May 2017).

Redman, R. 7-Eleven Pop-Up Store Serves Up Groceries to Dallas Hospital. Available online: https://www.supermarketnews.com/retail-financial/7-eleven-pop-store-serves-groceries-dallas-hospital (accessed on 5 November 2020).

ReGen Villages Holding. REGENVILLAGES. 2016. Available online: http://www.regenvillages.com (accessed on 18 May 2017).

Register, R. Ecocity Berkeley: Building Cities for a Healthy Future; North Atlantic Books: Berkeley, CA, USA, 1973.

Reid, C. Anne Hidalgo Reelected as Mayor of Paris Vowing to Remove Cars and Boost Bicycling and Walking. Available online: https://www.forbes.com/sites/carltonreid/2020/06/28/anne-hidalgo-reelected-as-mayor-of-paris-vowing-to-remove-cars-and-boost-bicycling-and-walking/?sh=ba645d11c852 (accessed on 5 November 2020).

Reid, C. Anne Hidalgo to Make Good on Pledge to Remove Half of City's Car Parking Spaces. Available online: https://www.forbes.com/sites/carltonreid/2020/10/20/paris-mayor-anne-hidalgo-to-make-good-on-pledge-to-remove-half-of-citys-car-parking-spaces/?sh=423352b916ec (accessed on 10 November 2020).

Reimer, J. The 15-Minute Infrastructure Trend That Could Change Public Transit as We Know It. Available online: https://360.here.com/15-minute-cities-infrastructure (accessed on 5 November 2020).

Reinartz, W.; Wiegand, N.; Imschloss, M. The impact of digital transformation on the retailing value chain. Int. J. Res. Mark. 2019, 36, 350–366.

Research and Market. Size is Expected to Grow from $410.8 Billion in 2020 to $8201.7 Billion. Available online: https://www.globenewswire.com/news-release/2020/10/05/2103315/0/en/Smart-Cities-Market-Report-2020-Global-Forecast-to-2025-Market-Size-is-Expected-to-Grow-from-410-8-Billion-in-2020-to-820-7-Billion.html (accessed on 12 November 2020).

Richards, G. Cultural tourism: A review of recent research and trends. J. Hosp. Tour. Manag. 2018, 36, 12–21.

Rittel, H.; Webber, M. Dilemmas in a general theory of planning. Policy Sci. 1973, 4, 155–169, Reprinted in Developments in Design Methodology; Cross, N., Ed.; Wiley: Chichester, UK, 1984; pp. 135–144.

Rodríguez-Pose, A.; von Berlepsch, V. Does population diversity matter for economic development in the very long term? Historic migration, diversity and county wealth in the US. Eur. J. Popul. 2019, 35, 873–911.

Roggema, R. Swarm Planning: The Development of a Spatial Planning Methodology to Deal with Climate Adaptation. Ph.D. Thesis, Delft University of Technology, Delft, The Netherlands, 2012.

Roggema, R. The future of sustainable urbanism: A redefinition. City Territ Archit. 2016, 3.

Roggema, R. Three urbanisms in one city: Accommodating the paces of change. J. Environ. Prot. 2015, 6, 946–956.

Roggema, R. Towards enhanced resilience in city design: A proposition. Land 2014, 3, 460–481.

Roggema, R. The Design Charrette: Ways to Envision Sustainable Futures; Springer: Dordrecht, The Netherlands; Heidelberg, Germany; London, UK, 2013; p. 335.

Roggema, R.; King, L.; van den Dobbelsteen, A. Towards Zero-Carbon metropolitan regions: The example of Sydney. Energies 2017. forthcoming.

Roggema, R.; Martin, J.; Arcari, P.; Clune, S.; Horne, R. Design-Led Decision Support Process and Engagement; VCCCAR: Melbourne, Australia, 2013.

Roggema, R.; Martin, J.; Vos, L. Governance of Climate Adaptation in Australia: Design Charrettes as Creative Tool for Participatory Action Research. In Action Research for Climate Change Adaptation: Developing and Applying Knowledge for Governance; Van Buuren, A., Eshuis, J., van Vliet, M., Eds.; Routledge: London, UK, 2014; pp. 92–108.

Roggema, R.; Thomas, L. Critical mapping for transformational cities. In Proceedings of the Passive and Low Energy Architecture (PLEA 2017), Edinburgh, UK, 3–5 July 2017.

Roggema, R.; Vos, L.; Martin, J. Resourcing local communities for climate adaptive designs in Victoria, Australia. Chin. J. Popul. Resour. Manag. 2014, 12, 210–226.

Rong, B.; Han, S.; Kadoch, M.; Chen, X.; Jara, A. Integration of 5g networks and internet of things for future smart city. Wirel. Commun. Mob. Comput. 2020, 2020, 2903525.

Roselund, M. Dimensions of the eco-city. Cities 1997, 14, 197–202.

Russell, J.S. The Agile City. Building Well-Being and Wealth in an Era of Climate Change; Island Press: Washington, DC, USA, 2011.

Ryan, C. Eco-acupuncture: Designing and facilitating pathways for urban transformation, for a resilient low-carbon future. J. Clean. Prod. 2013, 50, 189–199.

Ryan, C.; Biggs, C.; Trudgeon, M. Eco-acupuncture and greenaissance: Designing urban interventions for resilient post-carbonaceous futures, from Victoria (Australia) to Florence (Italy). In Proceedings of the State of Australian Cities Conference, Sydney, Australia, 26–29 November 2013.

Sala, G.; Aboutaleb, A.; Aki-Sawyerr, Y.; Cantrell, L.; Capp, S.; Durkan, J.; Kam-sing, W.; Medina, F.; Plante, V.; Calle, D.Q.; et al. C40 Mayors' Agenda for a Green and Just Recovery; C40 Cities: Rio de Janeiro, Brasil, 2020; pp. 1–43.

Sale, K. Dwellers in the Land: The Bioregional Vision; Sierra Club: San Francisco, CA, USA, 1985.

Salingaros, N.A. Compact city replaces sprawl. In Crossover: Architecture, Urbanism, Technology; 010 Publishers: Rotterdam, The Netherlands, 2006; pp. 100–115.

Salingaros, N.A. Complexity and urban coherence. J. Urban Des. 2000, 5, 291–316.

Sanders, E.B.N.; Stappers, P.J. Co-creation and the new landscape of design. CoDesign 2008, 4, 5–18.

Schwedhelm, A.; Li, W.; Harms, L.; Adriazol-Steil, C. Biking Provides a Critical Lifeline during the Coronavirus Crisis. Available online: https://www.wri.org/blog/2020/04/coronavirus-biking-critical-in-cities (accessed on 5 November 2020).

Shafie, F.A.; Dasimah, O.; Karuppannanb, S. Urban metabolism: A research methodology in urban planning and environmental assessment. In Proceedings of the 1st International Conference on Research Methodology for Built Environment and Engineering, Kuala Lumpur, Malaysia, 17–18 December 2013.

Shahab, S.; Allam, Z. Reducing transaction costs of tradable permit schemes using blockchain smart contracts. Growth Chang. 2020, 51, 302–308.

Sijmons, D. Het Casco-Concept, een Benaderingswijze Voor de Landschapsplanning; Ministerie van LNV, directie NBLF: Utrecht, The Netherlands, 1992.

Sinxadi, L.; Awuzie, B.O.; Haupt, T. Tackling spatial inequalities through mixed income housing: A qualitative analysis of stakeholder perceptions. In Proceedings of the 14th International Postgraduate Conference, Manchester, UK, 3–14 August 2020.

Sisson, P. How the '15-Minute City' Could Help Post-Pandemic Recovery. Available online: https://www.bloomberg.com/news/articles/2020-07-15/mayors-tout-the-15-minute-city-as-covid-recovery (accessed on 4 November 2020).

Sisson, P. What is a 15-Minute City? Available online: https://citymonitor.ai/environment/what-is-a-15-minute-city (accessed on 10 November 2020).

Smith, M. City to Test Road Closures, Putting More Space between Pedestrians. Available online: https://calgaryherald.com/news/local-news/city-to-test-weekend-road-closures-to-help-calgarians-keep-their-distance/ (accessed on 5 November 2020).

Stamatiadis, N.; Pappalardo, G.; Cafiso, S. In Use of technology to improve bicycle mobility in smart cities. In Proceedings of the 5th IEEE International Conference on Models and Technologies for Intelligent Transportation Systems (MT-ITS), Napoli, Italy, 26–28 June 2017; pp. 86–91.

Stevens, P. Cura Shipping Container Icus Open in Turin to Combat Covid-19. Available online: https://www.designboom.com/architecture/cura-shipping-container-icus-turin-covid-19-04-21-2020/ (accessed on 5 November 2020).

Sullivan, L. The Tall Office Building Artistically Considered; Lippincott's Monthly Magazine: New York, NY, USA, 1896.

Surowiecki, J. The Wisdom of Crowds: Why the Many Are Smarter Than the Few; Little Brown: Boston, MA, USA, 2004.

Swanson, C. Denver to Close Several Streets Starting Saturday to Allow for More Outdoor Activities. Available online: https://www.denverpost.com/2020/04/03/denver-streets-closed-coronavirus-covid/ (accessed on 5 November 2020).

Taleb, N.N. Antifragile—Things That Gain from Disorder; Random House: New York, NY, USA, 2012.

Teeuw, P. Duurzame Ideeën % DCBA Methodiek: Ambitie Stellen Volgens de Viervarianten-Methode; AEneas, Boxtel: Hertogenbosch, The Netherlands, 2010.

The Place Brand Observer. 31 Innovative Destination Marketing Campaigns and Place Branding Strategies to Serve as Inspiration in 2018. Available online: https://placebrandobserver.com/31-inspiring-destination-marketing-campaigns-place-branding-examples/ (accessed on 3 July 2019).

Tillie, N.; Van Den Dobbelsteen, A.; Doepel, D.; Joubert, M.; De Jager, W.; Mayenburg, D. Towards CO2 Neutral Urban Planning: Presenting the Rotterdam Energy Approach and Planning (REAP). J. Green Build. 2009, 4, 103–112.

Tjallingii, S. Planning with water and traffic networks. Carrying structures of the urban landscape. Res. Urban Ser. 2015, 3, 57–80.

Tjallingii, S.P. Ecopolis: Strategies for Ecologically Sound Urban Development; Backhuys Publishers: Leiden, The Netherlands, 1995.

Toker, Z.; Pontikis, K. An inclusive and generative design process for sustainable urbanism: The case of pacoima. J. Urban. Int. Res. Placemaking Urban Sustain. 2011, 4, 57–80.

Tomásek, W. Die Stadt als Oekosystem; Ûberlegungen zum Vorentwurf Landschafsplan Köln (The city as ecosystem; considerations about the scheme of the Landscape design Cologne). Landsch. Stadt 1979, 11, 51–60.

Townsend, A.M. Smart Cities. Big Data, Civic Hackers, and the Quest for a New Utopia; W.W. Norton Inc.: New York, NY, USA, 2014.

UN-Habitat. Un-Habitat Key Message on Covid-19 and Public Space; UN-Habitat: Nairobi, Kenya, 2020.

United Nations. World Urbanization Prospects. The 2007 Revision; United Nations: New York, NY, USA, 2008.

Van den Dobbelsteen, A.; Jansen, S.; van Timmeren, A.; Roggema, R. Energy Potential Mapping—A systematic approach to sustainable regional planning based on climate change, local potentials and exergy. In Proceedings of the CIB World Congress, Cape Town, Africa, 14–18 May 2007.

Van den Dobbelsteen, A.; Wisse, K.; Doepel, D.; Tillie, N. REAP2-new concepts for the exchange of heat in cities. In Proceedings of the 4th CIB Conference Smart and Sustainable Built Environments, Sao Paulo, Brazil, 27–29 June 2012.

Van Schaick, J.; Klaasen, I.T. The Dutch layers approach to spatial planning and design: A fruitful planning tool or a temporary phenomenon? Eur. Plan. Stud. 2011, 19, 1775–1796.

Van Timmeren, A.; Henriquez, L.; Reynolds, A. Ubikquity & the Illuminated City; TU Delft: Delft, The Netherlands, 2015.

Van Vuren, T. The 30-minute city: Designing for access. Transp. Rev. 2020, 40, 685–686.

VROM-raad. De Hype Voorbij, Klimaatverandering als Structureel Ruimtelijk Vraagstuk; advies 060; VROM-raad: Den Haag, The Netherlands, 2007.

Waldheim, C. Landscape as Urbanism. In Landscape as Urbanism Reader; Waldheim, C., Ed.; Princeton Architectural Press: New York, NY, USA, 2006; pp. 21–34.

Waller, S. $39M Spent on Temporary Infrastructure in Scotland. Available online: https://www.cyclinguk.org/news/ps39m-spent-temporary-infrastructure-scotland (accessed on 10 November 2020).

Waterman, R.E. Integrated Coastal Policy via Building with Nature. Ph.D. Thesis, Delft University of Technology, Delft, The Netherlands, 2010.

Weng, M.; Ding, N.; Li, J.; Jin, X.; Xiao, H.; He, Z.; Su, S. The 15-minute walkable neighborhoods: Measurement, social inequalities and implications for building healthy communities in urban china. J. Transp. Health 2019, 13, 259–273.

Western Australian Planning Commission. Designing out Crime. Planning Guidelines; State of Western Australia: Perth, Australia, 2009.

WHO, World Health Organisation. The World Health Organisation Quality of Life (WHOQOL); World Health Organisation: Geneva, Switzerland, 2012.

Whyte, W.H. City: Rediscovering the Center; Doubleday: New York, NY, USA, 1990.

Wijffelaars, M. Spain's Struggle against Covid-19 Hampers Its Economic Recovery. Available online: https://economics.rabobank.com/publications/2020/october/spains-struggle-against-covid-19-hampers-its-economic-recovery/ (accessed on 5 November 2020).

Wise, E. A gradual awakening: Broadacre city and a new American agrarianism. Berkeley Plan J. 2013, 26, 133–149.

Wolman, A. The metabolism of cities. Sci. Am. 1965, 213, 179–190. [PubMed]

World Commission on Environment and Development (WCED). Our Common Future; Oxford University Press: New York, NY, USA, 1987.

Yang, L.; van Dam, K.H.; Majumdar, A.; Anvari, B.; Ochieng, W.Y.; Zhang, L. Integrated design of transport infrastructure and public spaces considering human behavior: A review of state-of-the-art methods and tools. Front. Archit. Res. 2019, 8, 429–453.

Yigitcanlar, T.; Velibeyogly, K.; Baum, S. (Eds.) Knowledge-Based Urban Development: Planning and Applications in the Information Era; IGI Global: Hershey, PA, USA, 2008.

Zhekova, D. Is This the Future of Green Public Spaces? Available online: https://www.departures.com/travel/maze-like-park-vienna-social-distancing (accessed on 5 November 2020).

INDEX